THE NUM AND BRITISH POLITICS

The NUM and British Politics
Volume 1: 1944-1968

ANDREW TAYLOR
University of Sheffield

LONDON AND NEW YORK

First published 2003 by Ashgate Publishing

2 Park Square, Milton Park, Abingdon, Oxon OX14 4RN
711 Third Avenue, New York, NY 10017, USA

Routledge is an imprint of the Taylor & Francis Group, an informa business

First issued in paperback 2016

Copyright © 2003 Andrew Taylor

The author has asserted his moral right under the Copyright, Designs and Patents Act, 1988, to be identified as the author of this work.

All rights reserved. No part of this book may be reprinted or reproduced or utilised in any form or by any electronic, mechanical, or other means, now known or hereafter invented, including photocopying and recording, or in any information storage or retrieval system, without permission in writing from the publishers.

Notice:
Product or corporate names may be trademarks or registered trademarks, and are used only for identification and explanation without intent to infringe.

British Library Cataloguing in Publication Data
Taylor, Andrew, 1954–
   The NUM and British Politics
   Vol. 1:1944-1968. – (Studies in Labour history)
   1. National union of Mineworkers – History 2. Coal miners – Labor unions – Great Britain – History 3. Labor unions – Great Britain – Political activity 4. Great Britain – Politics and government – 1945 –
   I. Title
   331.8'8122334'0941

Library of Congress Cataloging-in-Publication Data
Taylor, Andrew, 1954–
   The NUM and British politics / Andrew Taylor.
       p. cm.– (studies in labour history)
   Includes bibliographical references and index.
   Contents: v. 1. 1944-1968
   ISBN 0-7546-0690-2 (alk. paper)
       1. National Union of Mineworkers–History–20th century. 2. Coal miners–Labor unions–Great Britain–Political activity–History–20th century. 3. Great Britain–Politics and government–20th century. 4. Coal miners–Great Britain–History–20th century. 5. Labor movement–Great Britain–History–20th century. I. Title. II. Studies in labor history (Ashgate (Firm))

HD6668.M6152N3578 2003
331.88'122334'094109045–dc21

2003045236

ISBN 13: 978-0-7546-0690-1 (hbk)
ISBN 13: 978-1-138-26360-4 (pbk)

Contents

General Editor's Preface		*vi*
Preface		*vii*
Acknowledgements		*xi*
List of Abbreviations		*xii*
1	Creating the New Order	1
2	Inclusion or Integration?	33
3	The Politics of State Capitalism	81
4	Conservatives and the NUM	121
5	The Politics of Industrial Decline	175
6	The Miners and Mr Wilson's New Britain	213
Bibliography		*257*
Index		*263*

Studies in Labour History
General Editor's Preface

Labour history has often been a fertile area of history. Since the Second World War its best practioners – such as E.P. Thompson and E.J. Hobsbawm, both Presidents of the British Society for the Study of Labour History – have written works which have provoked fruitful and wide-ranging debates and further research, and which have influenced not only social history but history generally. These historians, and many others, have helped to widen labour history beyond the study of organised labour to labour generally, sometimes to industrial relations in particular, and most frequently to society and culture in national and comparative dimensions.

The assumptions and ideologies underpining much of the older labour history have been challenged by feminist and later by postmodernist and anti-Marxist thinking. These challenges have often led to thoughtful reappraisals, perhaps intellectual equivalents of coming to terms with a new post-Cold War political landscape.

By the end of the twentieth century, labour history had emerged reinvigorated and positive from much introspection and external criticism. Very few would wish to confine its scope to the study of organised labour. Yet, equally, few would wish now to write the existence and influence of organised labour out of nations' histories, any more than they would wish to ignore working-class lives and focus only on the upper echelons.

This series of books provides reassessments of broad themes of labour history as well as some more detailed studies arising from recent research. Most books are single-authored but there are also volumes of essays centred on important themes or periods, some arising from major conferences organised by the Society for the Study of Labour History. The series also includes studies of labour organisations, including international ones, as many of these are much in need of a modern reassessment.

Chris Wrigley
British Society for the Study of Labour History
University of Nottingham

Preface

This book is the first of two volumes which examine the place of the National Union of Mineworkers in post-war British politics up to 1995 when the coal industry was returned to the private sector. The first volume covers the period from the NUM's formation in 1944 to 1968 when it seemed that the NUM and the mineworkers, battered by round after round of pit closures, were about to leave the political stage.

In several respects the approach adopted in this book to its subject matter is unfashionable and indeed old fashioned. First, it is unambiguously institutional in its approach in that it focuses on the NUM, the NCB and government. This is justified on the pragmatic grounds that institutions matter, that they are the vehicles whereby interests are articulated and that institutions encapsulate the processes whereby these interests are aggregated into behaviour. Second, the book concentrates on the political processes whereby the fragmentation and differentiation characteristic of British mineworkers were overcome and aggregated into a more-or-less coherent strategy. Third, the book's central focus is with politics and in particular high-politics. Its focus is the political behaviour of the NUM, how it interpreted its political environment, and the strategies it chose in response to that environment and in its relations with government and the NCB. This means that little space is devoted to the NUM's relationship with the Labour Party and the union movement except where these impinge directly on the NUM's relationship with high politics. More attention is devoted to the NUM's relationship with Conservative governments in the 1950s as these governments had a major impact on the evolution of the industry and the NUM.

In recent years writing on mineworkers has tended to emphasise the variety of the mineworkers' experience and the obstacles to the emergence of class consciousness and solidarity. This was a welcome and necessary corrective to a more 'heroic' style which tended to emphasise the mineworkers role as the vanguard of the proletariat, but it has two weaknesses. The first is that mineworkers often saw themselves as amongst the most conscious and solidaristic section of the industrial working class; and second, the emphasis on differentiation and fragmentation has obscured the efforts made to overcome these tendencies within the NUM. That these tendencies – solidarity and fragmentation – and the tensions they generated proved to be ultimately insurmountable does not detract from the need to explain and analyse the politics generated. This is why there is no detailed consideration in this book of politics within coalfields or with the complexities of day-to-day pit politics; the focus is on the mechanisms and structures which were created consciously to encourage mineworkers' solidarity and consciousness.

The NUM was the creation of, and articulated, a particular view of what it was to be a mineworker in mid-twentieth century Britain and this volume examines the consequent evolution of a *collective* identity in post-war politics. This identity was the product of the crises of the inter-war years and the political legacy of 1926 which dominates NUM politics in the period covered by this book. This legacy embraced a determination that the mineworkers should never again confront the state but should strive to influence policy via liberal democratic political means, and that nationalisation offered the only hope both for the mineworkers and the industry. Nationalisation achieved an iconic status but mineworkers were realistic enough to appreciate the limitations of what had been achieved in 1946-47, and the post-war politics of the NUM were dominated by the clash between the myth and reality of public ownership. NUM politics were determined both by the creation of nationalisation and by its degeneration as a political project into a technocratic strategy of industrial rationalisation. The latter cruelly exposed the NUM's inability and, at times, outright refusal to adapt to these changes.

To understand the NUM's politics we have to understand the context in which the NUM was created. Chapter 1 examines the 1944 reorganisation of the Mineworkers' Federation of Great Britain (MFGB) and the creation of the NUM. The purpose of the 1944 Rulebook was to find a constitutional order and political process which both expressed and controlled the centrifugal forces which were ever present in union and coal politics. The chapter also considers the wartime consensus that something had to be done about coal; as a result of the 1945 general election that something was nationalisation. Chapter 2 explores the early post-war years and the political implications of the country's almost total dependence on coal and therefore, the NUM. Pay and production were the key political–economic issues in these years but as important was ensuring that the mineworkers were not, as so often in the past, a source of turbulence in the British political economy. An important factor in avoiding this turbulence was the impact of the Cold War on the NUM's politics; this led to the creation of a right-wing majority determined to maintain support for nationalisation, orthodox Labour politics and the marginalisation of dissent. In these years the paradox was established that visibility and indispensibility did not produce a commensurate level of political influence. Chapter 3 examines this paradox by concentrating on the political nature of nationalisation and the NCB. The chapter is concerned with three questions; first, what was the purpose of nationalisation?; second, to what extent and in what ways did the NCB differ from the private owners?; and third, what was the role of the NUM in the new dispensation? These were questions asked by mineworkers with increasing frequency in the 1950s.

Chapter 4 examines what could have been an explosive relationship, that between the NUM and the Conservative governments of this period. In the event, the direct relationship between the two was minimal, which testifies to the NCB's importance as a buffer between the mineworkers and the government. Neither the NUM nor the Conservatives had any incentive to disturb the pattern of organisational and political relationships which had developed after 1947. Whatever its imperfections the status quo was accepted as the best obtainable and any changes were marginal. Nevertheless, Conservative ministers were alive to the NUM's potential power and sought ways to reduce the country's dependence on the NUM. Up to 1957 the pressure on the NUM was to facilitate the production of coal, thereafter the pressure was to cut output and rationalise production. Chapter 5 explores the reasons for, and the consequences of, this strategic shift in the pattern of fuel consumption in favour of oil. An important theme in all these chapters is the NUM's refusal to contemplate industrial action. Before 1956 this was because the NUM was too powerful and to do so would be irresponsible and contrary to the norms of nationalisation and democratic politics. Moreover, *pace* 1926, it would invite massive retaliation from the state. After 1957 the perception of many NUM leaders was that the NUM's power had evaporated and that the only realistic and responsible course open to the NUM was to co-operate in the rundown and secure the best possible deal for those who remained in the industry. In terms of political strategy this left the NUM with conventional lobbying and interest group tactics *via* Whitehall, the House of Commons, and the Labour Party. Chapter 6 examines the NUM's response to the rundown and how after a promising start, the NUM's expectation of a major policy shift by Harold Wilson's Labour government elected in 1964 was not merely disappointed but was replaced by an accelerated pit closure programme. The NUM's response was to acquiesce.

Permeating these chapters is a particular view of the relationship between organised labour and the state. The NUM's political relationships demonstrate how even the most potentially powerful union is politically subordinate to and excluded from key state decision making structures and processes. Despite its potential power, the NUM's political behaviour in this period was reactive and defensive with respect to governments of either party as well as the NCB. This volume shows that political power is not a commodity possessed but is a relationship determined by the matrix of material and ideological influences in which an organisation is located. The NUM's position in post-war politics is best described as *inclusion*. The NUM secured representative status and was consulted frequently albeit at arm's length by the NCB but was effectively excluded from core decision making.

Acknowledgements

The research on which this book is based has benefited over many years from the unstinting co-operation of officials and members of the National Union of Mineworkers. Successive officers at National and Area level have permitted me to range far and wide in the NUM's records and to use the material in my work on the mineworkers. I owe them and their members an enormous intellectual and personal debt. Although they cannot be listed I should also like to acknowledge the value of the innumerable conversations, interviews and correspondence with NUM members over the years.

I would also like to thank the Controllers of HM Stationery Office for permission to quote from various official publications and for permission to quote from the official papers deposited in the Public Records Office, Kew. Steven Bird at the National Museum of Labour History in Manchester found relevant material in the Labour Party archives. The chair of the Conservative Party granted permission to consult and quote from material in the Conservative Party archive at the Bodleian Library, Oxford. Other libraries and archivists were generous with their time and resources: the library at the National Mining Museum at Caphouse near Wakefield, the Durham County Record Office, the South Wales Miners' Library at the University of Swansea, and the Mass Observation Archive at the University of Sussex. The university libraries of Huddersfield, Sheffield, Leeds, and Manchester were used during the research as were the reference libraries at Huddersfield, Doncaster and Sheffield. Every effort has been made to trace all the copyright-holders, but if any have been inadvertently overlooked the publishers will be pleased to make the necessary arrangement at the first opportunity.

The bulk of the research and writing was completed while I was still a member of the Politics team at the University of Huddersfield and I would like to express my appreciation and gratitude to my former colleagues who provided such a congenial working environment. Many of the ideas in this book were tried out on successive groups of students who took my course, the Politics of Coal, their role as guinea-pigs cannot go unacknowledged. Dr Arthur Mawby, my friend and former teacher read the manuscript and made a number of valuable suggestions about the text. Alexander and Helena, as usual, provided a much needed element of scepticism about research and writing. Last, but by no means least, Georgina has been a consistent and constant source of encouragement and love. *La lucha continua.*

List of Abbreviations

ACP	Accelerated Closure Programme
AEA	Atomic Energy Authority
AGR	Advanced Gas-Cooled Reactor
CEGB	Central Electricity Generating Board
CINA	Coal Industry Nationalisation Act
CINCC	Coal Industry National Consultative Committee
CLP	Constituency Labour Party
CPGB	Communist Party of Great Britain
CPPI	Council for Productivity, Prices and Incomes
CRD	Conservative Research Department
DEA	Department of Economic Affairs
EHA	Extended Hours Agreement
FDWA	Five Day Week Agreement
GW	Gigawatt
JNNC	Joint National Negotiating Committee
JSCC	Joint Standing Consultative Committee
MAGB	Mining Association of Great Britain
MAGNOX	Magnesium Oxide Reactor
MFGB	Mineworkers' Federation of Great Britain
MLNS	Ministry of Labour and National Service
MP	Member of Parliament
NACODS	National Association of Colliery Overmen, Deputies and Shotfirers
NBPI	National Board for Prices and Incomes
NCB	National Coal Board
NCC	National Consultative Committee
NEC	National Executive Committee
NEDC	National Economic Development Council
NIC	National Incomes Commission
NPLA	National Power Loading Agreement
NRT	National Reference Tribunal
NUM	National Union of Mineworkers
PWR	Pressurised Water-Cooled Reactor
TUC	Trades Union Congress
WFTU	World Federation of Trade Unions

Chapter 1

Creating the New Order

Introduction

This chapter examines the creation of the National Union of Mineworkers in 1944 and the constitution under which it was to operate in the post-war period. It then examines the various schemes for the industry's future put forward during the Second World War, all of which recognised the inevitability of state reorganisation of the coal industry. Whether or not the industry remained in private hands depended on the result of the general election held in 1945.

Reorganisation: The National Union of Mineworkers

The NUM and its predecessor, the Mineworkers' Federation of Great Britain (MFGB), were federal unions and the NUM displayed many of the strengths and weaknesses associated with a federal political system, such as the need to promote unity, aggregate diverse interests, and manage centre-periphery relations. The study of internal union politics has tended to concentrate on factional struggle and the role of the union 'boss' to the exclusion of the union's rulebook. The rulebook is analogous to a codified constitution, providing the ground rules and framework for the exercise of power and authority, as well the boundaries to the exercise of power and authority. However, a political system cannot be understood solely by reference to the constitution whose operation is conditioned by history, myth, political culture, the attitudes of those who make the system work, and the wider environment.

The MFGB's national bureaucracy was limited (only in 1919 was the general secretary's post made full-time), the Executive was made up of members drawn from the largest coalfields with a rota from the smaller ones, and its members owed their primary loyalty to the district unions. The supreme policy making body was the annual Conference made up of district delegations composed of full time and lay officers as well as ordinary members in proportion to district membership. Special conferences could be called to deal with specific issues or immediate crises, or to pronounce on national matters such as industrial action. The district unions were the powerhouses of the MFGB:

> The consequence was that the Federation often seemed more a collection of disparate unions than a national body. Frequently men sat on its executive or in its conference as ambassadors from coalfields rather than as participants in a collective enterprise. Coalfield chauvinisms emerged at moments of stress. Each district had its own history – usually longer than the Federation's – its own myths and symbols, and ethos. Against these fissiparous tendencies there was the fact of Federation, an institutional acknowledgement of the desirability of national organisation and action. Such aspirations had to combat not just district identities, but diverse economic conditions and expectations. Maintenance of the MFGB's unity was an art built around the recognition yet limitation of difference.[1]

The problems of the inter-war years, especially the defeat of 1926 and its consequences, reawakened interest in organisational reform and the MFGB's wage claim of 1935-36 demonstrated unequivocally that national unity behind a common strategy could deliver real gains. The experience of the wage claim also showed how dependent success was on the quality of the national leadership of the MFGB and the ability of the General Secretary and President to win and retain the confidence and support of the district unions. The internal politics of the MFGB and the quest for unity was further complicated by the secession of George Spencer and a majority of the Nottinghamshire miners in 1926. Although the MFGB and the Spencer union were united in 1937, this was achieved on the basis of mutual agreement and Spencer remained personally hostile to the creation of one union.[2]

Proposals for re-organising the Federation dated back to *The Miners' Next Step* (1912) and beyond but the MFGB's revival in the late 1930s led directly to the creation of a Reorganisation Sub-Committee (1937). However, the districts were not prepared to grasp the nettle of reorganisation and a Special Conference to discuss reorganisation was postponed twice. The difficulty was that a lack of unanimous agreement might have meant some districts being outside the reformed MFGB. Yet, by 1940 reform was back firmly on the agenda and a climate supportive of reorganisation was encouraged by the difficulties of mobilising the industry for war. So the NUM was in part a product of the war-time production crisis which increased the importance of the national level in the politics of the MFGB. The centralisation of production policy and collective bargaining entailed the centralisation of problem-solving and conflict resolution but this contrasted with the historical and cultural legacy of the MFGB.[3]

The 1944 Rulebook

The process which was to produce the NUM in 1944 began with Resolution 11 at the 1942 MFGB Conference sponsored by the Executive. In moving the Resolution, James Bowman, the Vice President, rejected 'splendid district isolation' declaring unity to be essential as the Federation's current organisation was incapable of responding to the industry's growing industrial and political centralisation.[4] In June 1942 the Executive had determined that reorganisation would be based on the centralisation of industrial and bargaining activity coupled with the minimum interference with the districts' internal practices. Some districts (for example, Yorkshire) favoured a speedy transition to a national industrial union by sweeping away the MFGB and replacing it with a more rational structure. This rationalism had permeated much of the left's thinking on reorganisation from 1912 onwards but the political impossibility of such a fundamental restructuring of the MFGB reinforced the Executive's conviction that one union could emerge only on the basis of district consent which depended on the preservation of district autonomy. A further difficulty with the rational solution was that it was identified (unfairly) with the Communists in the union; hence the political symbolism of Arthur Horner (a Communist from South Wales) and Sam Watson (a key Labour Party figure in the MFGB from Durham) working *together* on the reorganisation. The process of creating a consensus behind the new union required careful lobbying by the MFGB's National Officials and their supporters in the districts. Their argument was that the NUM's structure was both a recognition and reflection of the changes brought about by the war and an extension of the natural evolution of the MFGB's national role, which had been revitalised with the 1935-36 wage claim and the formation of the Joint Standing Consultative Committee (JSCC), and with the unification campaign at Haworth in 1937.

Critical to the stability and viability of the new union was the attitude of George Spencer and Nottinghamshire. Spencer was verging on retirement and after reunification in 1937 a new generation of district officials was emerging who were less concerned with the battles of the past but Spencer remained a powerful voice in favour of district autonomy and his arguments evoked a sympathetic response in many coalfields. Spencer recognised he was fighting a losing battle but, nevertheless, he described the Executive's proposals as 'a hybrid conglomeration of nonsense...You are leaving under the present circumstances a variety of organisations which are different in their functions, different in their intentions, different in the amounts that they are receiving, different in the amount of benefits they are going to pay. Can anyone say that that is one organisation? No; it is the negation of one organisation.'[5] The Executive's pragmatic response to these forceful criticisms was that they had to work with what existed. Autonomy would be chipped away as districts grew accustomed to working within the new union especially after the Model Rules had been approved and in any event, national rules would take precedence over district rules. The centre-piece was a National Executive Committee (NEC) responsible for the Union as a whole, and 'the activities of area representatives must be subject to the overriding authority of the national organisation.'[6] These principles were circulated to the districts and approved by the 1943 Conference, after which a draft scheme and rulebook were prepared and sent out for extensive consultation. The National Officials held meetings in every district and were painstaking in their preparations for the Special Conference called to approve the new constitution. The agenda was sent to the districts in February 1944 and, with obvious symbolism, the Conference was held in Nottingham in August.

The first serious controversy at the Special Conference came over the payment of capitation fees by the districts (£1 per member) which meant a substantial transfer of resources (some £600,000) to the National Union. These resources signalled a power shift and the reduction of district autonomy. Bowman and the Executive regarded the capitation fees as the 'acid test' because the NUM could not begin to function without its own funds; 'Either we believe in one Mineworkers' Union or we do not', Bowman said.[7] Bill Allen from Northumberland spoke for the majority of delegates when he argued that unless the capitation fee was paid 'we have been indulging in farce', and warned that unless the MFGB adapted the mineworkers would find themselves at a disadvantage in the post-war world.[8] An amendment sponsored by South Wales to postpone payment was lost by 204 to 405, ironically South Wales' main supporter was George Spencer.

The main business of the Conference was to approve a set of rules by which the NUM was to be conducted. These became the 1944 Rulebook which was based on four principles. First, the changing nature of the industry and the inevitability of a major post-war reconstruction underscored the need for a more centralised union but one which took into full account district interests and sensibilities.⁹ The consequent accommodation under Rule 6 of district interests (especially, Nottinghamshire's) led to criticism. Yorkshire, for example, proposed the amendment of Rule 6 to create nine new geographical areas rather than preserve the historic structure. Bowman conceded that the retention of the historic coalfield structure was a weakness, but argued that the rules allowed the structure to evolve, and that precipitate action would undermine the union; thus the Executive were 'seeking at this early stage to prevent any rupture in the formation of the [NUM]...'.¹⁰ He was obviously hoping to placate sentiments such as those of Lancashire, which proposed a decentralising amendment to Rule 7 which would have made amalgamations possible only on the basis of full agreement between the Areas affected; this would have limited the power to promote amalgamations which Bowman argued was a key aspect of the Executive's authority. The Lancashire amendment was supported by Spencer and Nottinghamshire on the grounds that 'the more you narrow responsibility and freedom, the more you undermine in the long run the permanency of the structure and the power.' The amendment was, however, defeated in a card vote by 243 to 266.¹¹

Second, the National Union could only work effectively if its policy embraced and expressed district interests which were moulded by a national Conference into a national policy. Conference would remain the supreme policy making body, with policy flowing up from the districts to Conference to be aggregated, before being prioritised and implemented by the National Executive. This created tension between Conference and NEC. Amendments to the rule relating to calling a Special Conference, for example, show how the Executive strove to balance national and district interests whilst avoiding the dominance of the union by one interest or group of interests. Thus, when Cumberland proposed that any Area with 50,000+ members (in 1944 South Wales, Yorkshire, Scotland, and Durham) could call a Special Conference in an effort to promote rank-and-file control over the National Executive, Bowman rejected this. He feared that it would prevent the Executive from acting in the interests of the whole union and would transfer too much power back to the Areas via the Special Conference.

Third, in the interests of national unity and organisational coherence the union could not be run by simple majorities. Conference delegations were determined by the size of the area membership but they were subject to an upper limit, and when this principle was applied to coalfield representation in the NUM the effect was to limit the representation of the larger coalfields while over-representing the smaller. Running through the history of the MFGB was a tension between Executive authority and the ultimate sovereignty of the membership. Bowman argued that the NEC was capable of meeting the demands of both organisational efficiency and grassroots control, and thought it was erroneous 'to pretend that somebody who has gone to London as a National Executive member has suddenly lost all his district outlook and conceptions of class struggle as somebody who is now moving in the purple and does not understand coalfield psychology. The [Executive] is made up of men elected from the coalfield, and are now to be elected by ballot vote, to go to London, to conduct the affairs of the organisation...'. Bowman concluded that 'If they...have not your confidence, then when the ballot comes round, send someone who has.'[12]

Fourth, ultimate sovereignty resided in the membership who would elect the national officials and vote on questions such as industrial action but the Rulebook was designed to minimise the impact of the membership (or combinations of districts) on the National Executive. Under Rule 41 any Area dispute which seemed likely to result in a stoppage had to be reported to the NEC and no industrial action could be taken by an Area until the NEC had given its approval. Amendments from South Wales and Lancashire (later withdrawn) sought to keep approval of Area stoppages in the hands of the Areas but this was opposed by the Executive on the grounds that as the Areas would expect national support it was logical that the National Executive should give or withhold approval in the light of national policy and the interests of the union as a whole. Reproducing MFGB practice, Rule 43 stipulated that a two-thirds majority of those voting in a national ballot was required to sanction national industrial action. Cumberland proposed a simple majority but this was rejected as Rule 43 was 'dealing with one of the most important decisions... A national strike is something which should only be entered into after the most profound consideration, and after the rank and file have been fully consulted.'[13] The amendment was lost. Thus, the 1944 Rulebook was designed to create a national union by drawing away sufficient sovereignty from the district unions, but at the same time reconcile them to this loss by securing a balance between large and small coalfields, between left and right, through a political process designed to achieve consensus.

The NUM's Political Process

The political process generated by the Rulebook was, like the Rulebook itself, an attempt to manage the tension between 'pit' and 'union' politics in mining trade unionism.[14] *Union* politics were centripetal and reflected the mineworkers' goals as expressed by the NUM and its Areas; union politics tends towards cooperation, collaboration and compromise with the employer and the state in the interests of the combined membership whose material interests were enshrined within and expressed by, policy as determined by Conference. *Pit* politics were a centrifugal force. Politics at the point of production were an often volatile mixture of accommodation and conflict. The organisational and political linkage between union and pit politics was provided by branch officials who were therefore in a complicated position, sandwiched between the union and the membership. Their position was further complicated by the union's involvement in boosting output during the war and by the subsequent nationalisation. Local NUM officials were the legitimate representatives of NUM authority and were required to follow its policies, yet they owed their position to their election by local members and much of their authority depended on their standing with the local workforce and their ability to manage successfully pit industrial relations. The 1944 Rulebook tended to elevate union above pit politics which led to the relative exclusion of pit level discontent from the NUM's political process.

The NUM's Constituent Areas were autonomous, not sovereign. The Annual Conference made NUM policy but ultimate sovereignty lay with the membership. Conference was a small gathering of Area delegations in which rank and file delegates predominated, and decisions were taken on the basis of Area delegations voting as a block. A preliminary Conference agenda was presented to the Areas several weeks before Conference for discussion and the preparation of amendments. The final agenda was available not less than six weeks before Conference and there was considerable politicking between and within the Areas over the agenda. Where Areas submitted similar or identical resolutions attempts were made by national and area officials or by the powerful Conference Arrangements Committee to create a common platform by reconciling resolutions ('compositing') or securing the withdrawal of one in favour of the other. Compositing was an extremely important part of the NUM's political process.[15] The objective was the creation of the broadest possible area of agreement before the Conference began. The NEC was also given the power (subject to confirmation by Conference) to exclude any resolution which had been voted on during the previous two years or any resolution deemed by the NEC to be contrary to union Rules and Objects.

Delegations were given mandates by their Areas but these mandates can best be seen as guidelines for the delegation. Even when a mandate was clear (i.e. to support or oppose on a particular issue) there was room for delegation discretion because the delegation had to be able to negotiate within Conference with other parts of the NUM. The NUM Conference lasted at most a week and even though each Area was restricted to proposing three resolutions (plus amendments and composites), the agenda could be enormous. Many of the initial resolutions overlapped and the subtle differences between some of them had to be clarified. Compositing achieved these objectives yet, although crucial to the NUM's political process, was not mentioned in the Rules or Standing Orders. The importance of compositing was not simply procedural but political because it encouraged agreement without public argument and so helped to forge an intra-NUM consensus. Few composited were defeated. The basis of compositing was not necessarily securing a middle way as compositing sought not to exclude the extremes (though this happened) but to maximise agreement. Thus, it was possible to have a militant composite resolution if militancy was the mood of Conference. Compositing required a willingness to composite and refusal to composite invariably resulted in the defeat of the non-composited resolutions.

NEC advice and recommendations carried considerable weight in the NUM's political process and amongst delegates. The NEC strove to achieve unanimous agreement at Conference but if this was not possible it would mobilise support to defeat resolutions of which it disapproved. Its objective in agenda politics was to minimise conflict and maximise agreement. NEC support for a resolution generally meant it was passed; likewise if the NEC opposed a resolution it would usually fail. Should the NEC urge an Area delegation to withdraw a resolution it was usually accompanied by profuse expressions of sympathy for the resolution's objectives, but insistence that 'the facts of the situation, etc.' demanded that the resolution be withdrawn. The NEC could thereby indicate a predisposition to support a resolution in future when conditions permitted. If an Area delegation refused to withdraw the NEC would force a vote, but only if it was sure it could win as a public defeat for the NEC would erode its authority. Similarly, an Area delegation which moved a resolution against the NEC's wishes or which refused to composite did so in the knowledge that it would probably be defeated. Sometimes an Area would refuse to composite in order to force a debate on a contentious issue and so publicly register its dissent. The NEC could request Conference or an Area to remit a resolution to the NEC to permit the NEC to act upon it in its own time. Remission was a half-way house between rejection and acceptance as a remitted resolution was not thrown out but neither did it become policy.

It was for the Area moving a resolution to accept or reject the call for remission but if it refused the resolution was put to the vote with an NEC recommendation to reject. Remitted resolutions did often 'get lost' in an NEC sub-committee or stagnate until events neutralised the resolution's intent.

Once Conference had made its decisions implementation was the task of the NEC. Conference decisions formed the basis of NUM policy but how policy was realised was the NEC's responsibility. Implementation depended on the NEC's interpretation of the NUM's environment and its perception of the union's priorities. There was, therefore, a further stage before the final configuration of NUM policy was known in which resolutions were prioritised. This took place at the first NEC meeting after the Conference, with decisions about wages always taking precedence. There were, of course, ample opportunities for flexibility in the negotiations with the National Coal Board (NCB). The NEC was composed of the national officials (the President and General Secretary did not vote but wielded great influence) and lay members elected by the Areas who could be senior Area Officials or even national figures in their own right. The NEC was not national in the sense that it was elected by the whole NUM in a single election; rather, it was elected by the Areas and those elected had a two-year term. It was designed to represent sub-national interests and provided a forum for their reconciliation and aggregation. The NEC was an aggregation of three elements in the NUM: Conference, the Areas, and the National Officials. Other sources of inputs to the NEC's deliberations were the National Coal Board and government. Conference decisions provided the basis of union policy and a yardstick by which the Areas judged the NEC but the final priorities were the result of a complex interaction between these three elements. The NEC administered the NUM between conferences and performed duties laid down for it by Conference. The NEC represented mineworkers as a whole whilst the Areas represented the local interests of mineworkers. The NEC was thus far more than an executive body as it also made policy by determining priorities. It could also appeal directly to the membership in a national ballot.

At the 1944 Special Conference an unnamed delegate from South Wales asked a question that must have been in the minds of many delegates: 'Who is to be the boss of the organisation? Is it the President or the Secretary'? Bowman replied: 'Being a democratic organisation it has not any bosses. It has officials elected to carry out specific duties.'[16] The NUM had only two full-time national officials (the President and General

Secretary) elected by the whole membership. The Vice-President was nationally elected but was not full-time. The President's role was left deliberately vague. He chaired the NEC and all national Conferences and under the 1944 Rulebook was responsible for ensuring that the union's business was conducted according to the rules. This gave the President considerable influence as the authoritative interpreter of the Rulebook and as long as he enjoyed the support of the NEC, on which there was usually an in-built Presidential majority, he was in a pivotal position. Yet, on paper the General Secretary was more significant as he was responsible for correspondence, preparation of papers and agendas, and also played a key role in negotiations. In practice, a political balance was maintained by the tradition in the NUM that the left and right each returned one national official, and from 1936 until 1981 the right had the President and the left the General Secretary. General Secretaries were invariably described by the left as 'prisoners' of the right-wing majority on the NEC but at times of stress relations between the General Secretary and the NEC could become bitter and acrimonious. Nevertheless, General Secretaries were, it should be emphasised, the servants of the NEC and were required by rule (as were all other NUM officials and members) to further only National Union and NEC policy. Once a policy had been determined, NEC members were required to support that policy irrespective of their personal or Area preferences.

This was a robust and flexible Constitution which balanced the rights of the membership and needs of executive efficiency in an industry which was centralising rapidly. The 1944 settlement did not create a national industrial union but did create a framework which allowed disparate coalfields with differing traditions to work together within the nationalised coal industry. Recognising the diversity of the coalfields was the secret of the strength and flexibility of the NUM's constitution, but flexibility also facilitated stagnation. Factionalism in the post-war NUM grew after the outbreak of the Cold War and the identification of Communists in the NUM as potential threats to national security, and this damaged the ability of the Constitution to evolve. Areas were merged to reflect the industry's decline but the basic distribution of power set out in 1944 was preserved by Cold War politics which led the NUM right to isolate left influence. Left-right factionalism hindered the NUM's evolution by entrenching a conservatism which found it impossible to envisage any political or industrial strategy other than those of nationalisation and parliamentary politics. In due course, the 1944 Rulebook became a source of political frustration in the NUM as the power structure it expressed, dominated by a right-wing oligarchy, was, by the late 1960s, blamed for the NUM's failure to respond effectively to decline.[17]

The tension between pit and union politics, present from the outset in 1944, grew under public ownership and exploded in reaction to the accumulated frustrations caused by the run-down of the industry after 1957 and the mineworkers' treatment by the 1964-70 Labour government. These developments contributed to the growth of pit-level discontent in the NUM politics in the early-1970s which saw the NUM move decisively to the left. The ability of the 1944 Rulebook to keep the NUM's fissiparousness under control was tested to destruction in the political and economic crisis which culminated in the 1984-85 dispute. The key issue in the fragmentation of the NUM was Nottinghamshire Area's objection to the collapse of the political and ideological balance within the NUM in the late 1970s. The main opponent of unification in 1944, George Spencer, had warned that 'centralization' endangered long-term stability as 'the more you narrow [district] responsibility and freedom, the more you undermine in the long run the permanency of the structure and the power' of the national union.[18] An alternative view was that the NUM was insufficiently centralised.

Reorganisation: The Coal Industry

The Official History of the coal industry at war commented that 'No other major British industry carried so many unsolved problems into the war; none brought more out.'[19] The wartime debate on the industry's future produced a consensus that the industry's problems could be resolved only by state intervention. There were three sets of proposals on coal's future: the government's Reid Report, the Mining Association's Foot Plan, and the NUM's *The Miner's Case*. The political background to the Reid Report and the Foot Plan was the 1942 coal crisis which resulted in 'a new and powerful Ministry of Fuel and Power [which] had given a severe shock to the coal-owners.'[20] Naturally, the tendency is to see 1942 as the beginning of an inexorable process culminating in nationalisation, but in discussing pre-nationalisation coal politics it is vital to remember that the MFGB and then NUM may have hoped but did not expect that Labour would win a post-war election. The Executive recognised the political reality that the Coalition government would not sanction public ownership of the mines and it therefore expected (despite the frequent calls for public ownership) a state sponsored restructuring of the industry. This would have been the logical outcome of government policy in the 1930s.

The Reid Report

The industry's plethora of long- medium- and short-term problems merged in 1942 in a major production and political crisis which forced the Coalition government to create the Ministry of Fuel and Power and take control of the industry. Under the pressure of war, policy concentrated on the immediate goal of maximising output from the existing industry, but there was also widespread concern in government at the ability of the industry to deliver the coal that would be needed to fuel the post-war recovery.[21] In September 1944 the Ministry appointed a Technical Advisory Committee chaired by Sir Charles Reid, a mining engineer who was Director of Production at the Ministry of Fuel Power and had been General Manager of the Fife Coal Company. The Committee was to examine the production process from 'coal face to wagon, and to advise what technical changes are necessary in order to bring the Industry to a state of full technical efficiency.'[22]

The Committee was composed of seven mining engineers drawn from some of the largest coal combines and they stressed that their 'recommendations have been formulated from our professional viewpoint.'[23] Underpinning the Committee's assessment was a model for the post-war coal industry: the Nottinghamshire coalfield. The Reid Report argued that a high productivity industry could not be achieved solely by technical change but required a dramatic transformation of workforce attitudes and the creation of a culture of cooperation. The Reid Report's conclusion that drastic technical reorganisation was practicable and necessary was uncontroversial in the government and much of the industry. The difficulty was how to achieve this. The technical changes sought by the Reid Report were derived from a comparison of the inter-war performance of the British industry with German, Dutch, Polish, and to some extent, US coal mines. The Committee's main concern was the low level of productivity growth (defined as Output per Man Shift, OMS) in Britain compared to coal mining in other countries. Low productivity was blamed on the lack of modern underground haulage systems, the archaic layout of many mines, poor or non-existent training for mineworkers, the reluctance of mineworkers to make full use of new techniques and machinery because of a fear of unemployment, the production of coal in non-standard sizes and qualities, the exploitation of reserves with a view to a quick return rather than the maximisation of productive output, and a legacy of poor industrial relations. These problems were exacerbated by the industry's fragmented ownership, which made it impossible to secure agreement to close down or merge inefficient mines into larger units reconstructed to make optimum use of modern mining techniques and machinery, and which also made it difficult to raise capital for investment. Investor confidence was not helped by political uncertainty over the industry's future.

The Reid Report believed that the industry had managed to achieve the worst of all worlds:

> On the continent the existence of closely organised industries has encouraged the collective examination of their problems and a scientific and analytical assessment of future prospects. In the United States the attitude has been individualistic, with a vigorous enterprise, generally prepared to accept all the implications of the theory of the survival of the fittest, and to see the installation of the latest machinery as the proper corrective to impaired competitive power.[24]

British mine managers were wholly committed to neither approach and the absence of a clear strategy undermined attempts to modernise the industry. Fundamental to modernisation and efficient management were the mining engineers. British mining engineers enjoyed neither the status nor the technical freedom of their Continental counterparts and were not exposed to the ferocious competition in the United States which forced the adoption of radical technical solutions. The mining engineer's lack of status, the absence of incentives to innovate, and the industry's limited financial resources reduced the engineer's influence to that of ensuring the industry in its existing form continued to function.

The Reid Report noted that the industrial relations systems of Germany, Holland and the United States were very different from each other but were more effective than Britain's. Before the rise of the Nazis the German coal industry (like the British) had suffered from heavy unemployment and, at times, poor industrial relations, yet productivity had increased consistently. In Holland industrial relations were good with very few strikes, whereas in the US there were frequent and bitter disputes (often involving bloodshed) but the mineworkers had cooperated with management to achieve very high levels of output and therefore pay. The 1926 General Strike was a determining event in Britain, the Committee noted, because:

> the mineworkers were generally in no mood for willing co-operation with the employers. They mostly refused to recognise their wages, in the long run, must depend upon the progressive efficiency of the Industry. Mechanisation, though seldom countering active resistance, was generally received by the men with little enthusiasm...where machinery was installed, its potential savings seem largely to have been dissipated by a quiet but effective determination that the

number of men discharged should be kept as low as possible...they have steadfastly required the observance of old customs and traditions which are inappropriate to the conditions of mechanised mining, and have thus put a brake upon the modernisation of the Industry.[25]

The resentments engendered by the mineworkers' defeat in 1926, unemployment and short-time working inevitably encouraged minimal cooperation and low productivity, but the lack of cooperation and production problems during the war when the national interest called for the utmost cooperation, the Reid Report argued, indicated the intractability of these attitudes. There were exceptions to this gloomy picture which suggested the British coal industry could be regenerated. The exemplar was the Midlands coalfield whose easier geology undoubtedly helped its better productivity record, but geology alone could not explain all the differences. Management/workforce relations, the Reid Committee noted, played a crucial role in explaining the difference, as 'relations between the employers, the miners' leaders and the men, over a considerable period, have been good. It is clear that there has been a definite disposition on the part of the men to co-operate, and their leaders have encouraged them to produce to the limit.'[26] If a similar level of cooperation had been forthcoming in other coalfields, the Reid Report concluded, OMS in 1939 would have been substantially higher despite the industry's technical weaknesses.

Poor industrial relations and a lack of cooperation were major reasons why productivity failed to increased appreciably and 'the problem of securing full cooperation...is the most difficult that the Industry has to face.' The post-war success of the industry depended crucially on the attitudes of those working in the industry but attitudes on all sides were 'coloured by past history, future hopes, and the degree of trust placed in the honesty of purpose of the other party.' The report could offer no ready solution 'especially as there has been so long and unhappy a history of disputes, grievances and misunderstanding.'[27] The Reid Report did identify a number of preliminary steps which might form the basis for mutual understanding, one of which was that management and men should be brought into closer and more regular contact.

The Report proposed a considerable number of steps which could be taken to improve the mineworkers' situation. The mineworker should have a right to training, a decent wage, the redress of grievances, security of employment and improved safety and health. Training would produce not only a more productive and committed workforce but also an identity of interest between management and men as the latter became more familiar with the industry's wider difficulties and would encourage upward mobility into management. Good wages were vital for efficient production, workforce commitment and to recruit and retain labour, so mineworkers' wages should be comparable to those of other skilled workers. Speedy redress of grievances would improve productivity by giving the miner a stake in the industry and avoid costly stoppages. Much of the necessary machinery had been created during the war and experience had shown that pit level disputes could be resolved by good-will and without striking. Poor attitudes and work habits were a result of the insecurity of the 1920s and 1930s but the 1944 White Paper *Employment Policy* committed government to avoiding mass unemployment. Rapid technical change and industrial reorganisation would inevitably produce localised unemployment so government should retrain redundant mineworkers and ensure new industry located in the coalfields. Technical change would improve health and safety but would also cause new problems (e.g., machine mining would increase dust diseases and faster haulage could lead to more accidents away from the face) and the mineworker had to be convinced that health and safety would not be sacrificed in the search for output. Improved health and safety would help create a more contented and productive workforce.[28]

In return for these rights, the Report expected the mineworkers to agree to 'A Fair Day's Work.' Modern mining machinery was expensive and was effective only if utilised to the maximum through continuous production which in turn required a flexible workforce working full shifts and not just the set stint. Lightning strikes, voluntary absenteeism and poor work discipline undermined productivity and the Reid Committee was adamant that the mineworker had to recognise that these practices were not in his long-term interests. Strikes and poor work discipline eroded cooperation and commitment and the Report insisted it was the unions' duty to co-operate with management in minimising disruption to production by using grievance procedures. Mineworkers would have to abandon 'custom and practice' which was inappropriate in a modern coal industry and doing so was identified by Reid as the test of the mineworker's willingness to respond positively to the demands of the new industry as well as being vital to management's right to manage. Mineworkers would have to cooperate fully and unreservedly in all measures to increase productivity and would have to accept that increased wages could only be based on improvements in productivity not increases in output. Unless full cooperation and cultural change was achieved the impact of the technical transformation of the industry would be drastically reduced.[29]

Reid saw poor productivity as, in part, a consequence of the industry's organisation so the report could not avoid the question of coal's future structure. On this it concluded 'it is not enough simply to recommend technical changes which we believe to be fully practicable, when it is evident to us, as mining engineers, *that they cannot be satisfactorily carried through by the Industry organised as it is to-day.'* [30] Reid argued forcefully that because of the economy's dependence on coal, rapid and immediate technical change was needed to improve supplies and cut costs. During restructuring pits would close (many permanently) and methods of production would change. This could not be achieved painlessly and it would have political consequences. Reconstruction would be expensive and as the industry had been unable to provide the funds for capital investment before 1939, who would finance post-war reconstruction? Reid did not answer directly this question but the implication was clear. For most of the inter-war period the orthodox view was that there were too many small mines and reorganisation on the scale envisaged could only happen if the state became directly involved in reorganisation. This was all the more necessary because of the historic refusal of the private owners as a body to use the provisions of the Coal Mines Act (1930) and recommendations of the Coal Mines Reorganisation Commission to voluntarily restructure the industry. The Reid Report believed that:

> an Authority must be established which would have the duty of ensuring that the Industry is merged into units of such sizes as would provide the maximum advantages of planned production, of stimulating the preparation and execution of the broad plans of reorganisation made by these units, and of conserving the coal resources of the country. The existence of such an Authority, endowed by Parliament with really effective powers for these purposes, is, we are satisfied, a cardinal necessity.[31]

Coal would remain in the private sector but the state would compulsorily reorganise it into a few large combines and would determine its operational objectives and policy. Coal was clearly too important to be left to private enterprise, but the Report did not indicate what should be done if the coal owners resisted reorganisation.

The Reid Report was inevitably (and wrongly) seen as a general indictment of private ownership. The Report saw the industry as a prisoner of its past and this legacy was overwhelmingly negative. The Reid Report crystallised opinion in favour of an increased role for the state despite the problems of wartime control, by arguing that the national interest demanded economically priced supplies of coal, and that this must take primacy over the interests of both the mining companies and the workforce. The Reid Report bolstered the case for state intervention; indeed after Reid no one seriously doubted that the state had to play a crucial role in restructuring, that technical change on so huge a scale could not be divorced from organisational change and that both could be nullified by non-cooperation and antagonism from the workforce.

The Foot Plan

Despite a deserved reputation for myopia and intransigence the Mining Association of Great Britain (MAGB) was sufficiently politically sensitive to recognise that the 1942 crisis posed a considerable threat. In March 1942 Sir Evan Williams (MAGB President) told the Joint Standing Consultative Consultative Committee (JSCC), the forum where the MFGB and MAGB met to discuss issues of common interest, that he believed the time was now ripe to consider the industry's future and he proposed a joint inquiry with the mineworkers. This interested the MFGB but as they were as politically committed to public ownership as the owners were to private enterprise, the MFGB refused cooperation if the intention was change the union's policy. The MAGB accepted this but suggested that 'politics' were a matter for each body and political differences need not rule out a common approach on technical matters. On this basis the MFGB and MAGB appointed a joint sub-committee to prepare proposals for submission to the Ministry of Fuel and Power but in March 1943 the MFGB withdrew fearing that its cooperation would damage the case for public ownership of the mines.[32]

The MAGB had established its own post-war policy committee in September 1942. This functioned sporadically and after eighteen months concluded that the industry would have to be concentrated into fewer, larger undertakings. The Association was conscious of the collectivist drift of wartime politics and of public and governmental disquiet with their conduct of the industry's affairs. So to improve their image and ensure serious notice was taken of their proposals in May 1944 the MAGB appointed an Independent Chairman with full authority to formulate a post-war policy.[33] The chairman was Robert Foot, and under him the MAGB produced a plan for the industry's future, because Labour's presence in the Churchill Coalition government established in 1940 and the growth of the wartime state's intervention in the operation of the coal industry, worried the MAGB but equally they expected that control was a wartime expedient and would be followed as in 1921 by decontrol. The Association was also aware that Churchill opposed strongly extending wartime control over the industry into *de facto* public ownership, and the Conservative backbench MPs' defeat of Hugh Dalton's 1942 coal rationing scheme indicated considerable support for the continuation of private enterprise in the coal industry. However, the MAGB's proposals, known as the Foot Plan after its author, was an attempt not to avert state intervention but to influence the nature and extent of the state's role in the industry's affairs.

Foot identified three objectives for the industry: a good standard of living for mineworkers which 'implies continuous employment at a good rate of wages'; a reasonable return on capital 'in what always must be a relatively hazardous environment'; and the provision to the consumer of 'the quantities of coal desired at an economic price.' To achieve these objectives Foot called for radical change:

> From the miner, good and continuous work with flexibility and a liberal approach to mechanization and all cost-reducing processes. From the Management, enterprise, efficiency and, together with the miner, adaptability to all modern methods, technical and otherwise, leading to greater efficiency; and from the consumer an understanding of the immediate problems that face the Industry, and sufficient patience to give the miner and owner a chance to make their respective but at the same time mutual contributions.[34]

As in the Reid Report the basis of rejuvenation was cooperation but the Foot Plan did not question the continuation of private ownership. In response to Foot's own question: 'Are the present colliery owners the right men to be trusted with the control of the coal mining industry in the future?', Foot answered 'Yes, they are, provided (and this an essential proviso) they are prepared to accept the drastic kind of reorganisation which I feel the difficulties and seriousness of the present situation make imperative.'[35] Foot acknowledged the industry's heavy burden of history but believed the industry had changed for the better in recent years and this change should be extended, not swept away. To this end investment in safety, housing and training would improve management-worker relations. Government control, which was far more extensive than during the 1914-1918 war, had not transformed the industry and this did not give confidence 'that nationalisation, or public ownership in some shape or form, holds out a better hope of the solution of the long-term problem of the Industry than can be offered by private enterprise.'[36]

To reconcile the miners to private ownership Foot placed at the centre of his plan a new ideology, National Service. This was to be industry's governing principle and was neither a negation of private enterprise nor support for state imposed organisation. It required that all colliery owners should 'accept a joint responsibility for the efficient management of the Industry, and should be prepared to create and operate effective organisations nationally and in the districts to enable this responsibility to be properly discharged.'[37] Owners had to accept the primacy of the national over the sectional, and accept that they were trustees of the industry for the nation and behave accordingly by reorganising the industry around National Service.

To achieve this Foot set out Twenty Principles. The Twenty Principles called for industry to be treated as a single entity rather than an assemblage of independent districts and companies, though within the single industry each enterprise would be self-governing and autonomous. Each enterprise would be required to operate according to the best available practice, to produce coal economically and with due regard to conservation and safety. Enterprises would utilise the most up-to-date technology and practices with the object of bringing the industry to the highest state of efficiency. The industry would be encouraged to form larger enterprises 'by agreement if possible, by compulsion in certain circumstances.' To encourage workforce cooperation no pit would be closed as a result of reorganisation until a new mine was opened or arrangements had been made for the speedy transfer of redundant labour to new coalfields or non-mining jobs. Price competition, which had been used to force down wages, would be eradicated by central selling, mineworkers would be guaranteed employment security and underground wages would be at the top of the wages league. Capital should enjoy a 'reasonable' return, yet every effort would be made to encourage mineworkers to identify with the industry and they would be 'given every opportunity to make their full contribution towards its greater prosperity.' Liaison between the two sides would be established at every level of the industry and close relations with the Ministry of Fuel and Power would be established. Collective bargaining would be maintained and the best of modern personnel practices would be utilised to ensure the workforce's commitment to the industry. This would be supplemented by better training and education for mineworkers.[38] The industry would be controlled by a Central Coal Board (CCB), legally constituted by Parliament as the trustee for the nation and responsible for ensuring the industry adhered to the ideals of National Service. The CCB would have a full-time chairman and no more than 15 members who would be 'men of high standing and good reputation actively and presently engaged in its work who can be relied upon to approach their responsibilities in a statesmanlike way and to think nationally and not sectionally.' The CCB would control all undertakings employing more than 30 persons underground and its decisions would be binding, and its chairman would be appointed by the CCB.[39]

Foot's Plan was published in January 1945 but was damned by its origins. The Reid Report was perceived as the work of neutral experts whereas the Foot Plan was seen as a desperate attempt by a discredited group of employers to keep control of their assets. On 22 February 1945 the MAGB Central Committee announced that Foot's proposals enjoyed the industry's support and Foot would now draft the CCB Constitution and the Covenant to be entered into by each district association. These were published in April and Foot announced 'that the Industry had, to the extent of over 95 per cent of the output of the country, given its approval to the principles and structures recommended in his proposals and had assured him of its support in making them effective in the first instance and thereafter ensuring their maintenance.'[40] The Plan was described by *The Economist* as 'an honest attempt by an honest man to solve a difficult problem and the fact that it no way solves the problem does not detract from the sincerity of the attempt.'[41] *The Economist* agreed that state control was inferior to private enterprise but believed that the problem was that 'past experience provides plenty of occasion for doubting whether there is likely to be, in the particular case of the coal industry, such a confluence of public and private interest...' The primary obstacles, *The Economist* argued, were the industry's history and the owners' lack of credibility. After 1926 the owners had had the opportunity and powers necessary to restructure the industry but had failed to do so, which suggested the existing management 'is apparently helpless to put its own house in order.' Obloquy was heaped on the CCB which denied the miners, the consumers and the government a voice in the industry's affairs. The miners' exclusion was in itself sufficient to cripple the Plan, as whatever its other defects 'it is doubtful whether they would be satisfied with a scheme which gave them so little direct responsibility for the industry, much less, indeed, than they have under the wartime control scheme, which is based on tripartite responsibility of government, owners and miners.' *The Economist's* final verdict was devastating: 'it is difficult to resist the conclusion that Mr Foot's scheme would result in almost precisely the present state of affairs. There would be no effective movement towards closer integration – for at this time of day it is simply impossible to believe that voluntary methods...will achieve anything. There would be no drastic re-equipment... There would be no improvement in the labour position... it would be operated entirely by the mine-owners without any government Control.'[42]

The Miners and Nationalisation

The MFGB's energies were absorbed by wartime problems and creating the NUM, and its Executive therefore devoted little time to thinking about the future of the industry. The publication in January 1945 of the Foot Plan changed this. In February 1945 the NUM Executive appointed a sub-committee consisting of Abe Moffat (Scotland), Joe Hall (Yorkshire), and Arthur Horner (South Wales) to draft a response to Foot.[43] The sub-committee's first draft report argued the NUM must present a positive case for public ownership and was critical of the TUC and Labour Party (and by implication the leadership of the NUM) for approving the principle of public ownership but doing nothing to prepare detailed proposals.[44] Continued private ownership was rejected as the private coal industry

> has failed to maintain out of itself a real and progressive standard of living for all those engaged in it. It has failed to maintain a real economic level as distinct from a continuation of subsidies, direct or indirect. A problem faces the industry now that, unless something is done to reorganise the whole of it with a view to meeting its obligations out of its own resources, chaos looms on the post-war horizon.

The political obstacles facing the NUM were considerable: the Coalition had refused wartime nationalisation, the Conservatives remained opposed, no newspaper (other than Labour ones) supported public ownership, and the Foot Plan's suggestions to improve the industry's technical performance were very similar to those advocated by the MFGB over many years. The draft conceded that even if the industry was to remain privately owned huge technical change was inevitable and would require state involvement. The draft also stressed that the CCB would have executive power and noted that the NUM and government would be excluded from its deliberations; this was clearly unacceptable even though the centralisation represented by the CCB constituted a revolution in MAGB attitudes. The draft stated that the NUM had two options when making the case for public ownership: either it could present public ownership as the NUM's response to the MAGB's proposals, or alternatively it could present its ideas as a joint TUC/Labour Party policy which was the logical extension of wartime control; and choice between the two depended on whether or not public opinion and the government would accept that the private owners could run the industry successfully on the National Service principle. Finally, the draft reminded the Executive that the Reid Committee was about to report and was expected to recommend a massive state led programme of restructuring and capital investment – which implied that the state would demand a dominant voice in the post-war industry. The draft report was subject to lengthy and detailed criticism by the Executive, which considered that it was too tentative, and that it would be a potentially fatal political mistake for the NUM to endorse any solution other than public ownership. The Executive therefore instructed the sub-committee to re-convene and that the NUM's response should restate the case for public ownership of the coal industry.[45] The result was the sub-committee's final report, *The Miners' Case. An Answer to the Foot Plan*. This argued that 'the Foot Plan will not solve the crisis within the industry, but will only lead to a further intensification of the crisis, greater restriction of output and protection of vested interests at the expense of the miners and the general public.'[46]

The Miners' Case believed that the basic problem in the industry was the failure of the colliery owners ever to act other than in their own narrow, selfish interests. Recurring crises could be averted only 'if the existing pits are completely reorganised and have applied to them the most up-to-date methods and equipment...this can only be achieved by a radical re-organization of the whole industry.' Reorganisation would only work, however, if the discontents of the workforce were addressed and there was no evidence that the Foot Plan would do so. Only a government commitment to improve mineworkers' conditions would be acceptable and the Foot Plan had explicitly excluded government from the industry's affairs. A state licensed private monopoly was unacceptable to the NUM and moreover the CCB had no power to reorganise the industry even though voluntary reorganisation had palpably failed. The Foot Plan was proposing a system of 'ultra private enterprise' which would worsen industrial relations and so the only effective answer to the industry's problems was state ownership. The Executive resolved to approach the TUC and the Labour Party with a request for a tripartite committee to draw up a detailed plan for the nationalisation of the coal industry.[47] This decision lends credence to the view that when Labour entered government it was unprepared for nationalisation – which is correct in so far as no detailed blueprint existed but is incorrect in so far as the basic organisational principles and philosophy of coal nationalisation had been set out in the 1930s and these were well understood by the leaders of the Labour government and the NUM.

The technocratic model of public ownership adopted by the Attlee government after 1945 and associated with Herbert Morrison was the dominant model of public ownership in the Labour movement.[48] In 1929, for example, Philip Snowden declared that nationalisation would be 'through a public corporation controlled...by the best experts and businessmen' and the MFGB was told that coal would be nationalised only after the private industry had been reorganised and rationalised.[49] The Labour government's Coal Mines Act (1930) advanced state-sponsored rationalisation and technocratic management but proved ineffective as it contained no mechanism for overcoming the coal-owners' refusal to abandon their notorious individualism. The collapse of the Labour government in 1931 and the refusal of the industry's owners to reform re-ignited interest in public ownership in the Labour Movement and the crucial debates which were to influence the post-1945 public ownership programme, took place after 1931.

Labour Party theorising on public ownership was dominated by Morrison. In November 1929 the Cabinet approved his public board scheme for the reorganisation of London Transport and he later commented that this 'provided a blueprint on which all the designs for nationalisation of industries after 1945 were broadly based.'[50] Control would be exercised by boards of experts appointed by the responsible minister and expertise was to be the only criterion for appointment. Morrison did not exclude trade unionists or employees from the boards but the final test of appointment was individual capacity. He required Board members shed prior commitments and affiliations, including trade union appointments. He thus considered Board members appointed directly by the unions to be undesirable as they would have to act in the industry's wider interests which might conflict with their union's, so that 'Within a year the Trade Union delegate will be regarded by the rank and file as a man who has gone over to the boss-class and cannot be trusted anymore.'[51]

This model did not go unchallenged and there was considerable support in some unions for worker control in publicly owned industries. However, Morrison argued that an active role for workers (or unions) in management was impractical. First, a significant degree of worker or union representation would reduce the board's efficiency and integrity as such members would have divided loyalties, whereas the technocrats' loyalty would be to the industry rather than to any sectional interest. Where workers were appointed, Morrison argued they should be appointed on equal terms with the other members, not just because of their social origins or political affiliations. Finally, Morrison argued that the worker was far more interested in his union and his immediate work situation and, as such, could not envisage the interests of the industry as a whole. The worker's interest was, inevitably, sectional and the first aim of public ownership was to improve the immediate work situation and the workers' level of cultural development. Only gradually would the worker penetrate the managerial sphere and public ownership's purpose was to create a new type of worker but the closer relationship between management and the worker could not be permitted to undermine management's right to manage. Despite public ownership, retention by the unions of their traditional functions of collective bargaining and representation, as well as the right to strike, indicated the continuation of basic conflicts of interest as power relations between manager and worker would be unchanged.[52] The Labour Party and the TUC worked jointly on schemes of public ownership and both accepted the public board model. The TUC General Council Economic Committee report *The Public Control and Regulation of Industry and Trade*, argued that details should remain vague but that a basic scheme could be formulated based on a controlling board of publicly appointed experts and that those employed in the industry would have no statutory right to representation on the board. The report was accepted reluctantly by the 1932 TUC.[53] By 1933 a party-union consensus had emerged and despite strong residual sentiments in favour of statutory worker representation the TUC accepted that management had to be expert and efficient but that this did not prevent the unions from having an influence over the composition of the board, or even having union officers appointed to the board, and that the unions had a right to claim a share in control.[54]

The MFGB was on the periphery of this debate. As a result of a debate at the 1932 Labour Party Conference the MFGB agreed to allow the Party and TUC to formulate public ownership proposals and in effect agreed to accept any scheme which emerge from these discussions.[55] The MFGB initially regarded the composition of the controlling public board as pivotal as it would not only lead to the transformation of the coal industry but also be a basic building block of a socialist society. Without giving the worker some role in the control of the industry there could be no real transformation as a failure to alter management structures would ensure the perpetuation of old patterns of control and attitudes.[56] The MFGB did concede the primacy of managerial technical expertise in a publicly owned coal industry but argued this should not be used as an excuse to exclude the mineworker (who was *the* expert in winning coal) from decision making at all levels in the industry. Ebby Edwards, the MFGB secretary, wrote prophetically that if the miners were excluded from managerial authority 'then inevitably there must grow up again all the old antagonisms between management and labour...a state of affairs which should be avoided at all costs.'[57] By the mid-1930s the MFGB's strategy shifted to criticising the coal industry's inefficiency and to arguing that public ownership would, by definition, be more efficient than private ownership. The emphasis on efficiency and rationalisation enabled the MFGB to identify public ownership with the national, as well as the miners', interest suggests a predisposition to accept the party's emphasis on community of interest and technocratic ability as well as a pragmatic recognition that the mineworker would have a subordinate role in management.

By mid-1936 the TUC's proposals for a publicly owned coal industry were made available to the MFGB and the Executive accepted the TUC's proposals and thereby committed themselves to the Morrisonian public board.[58] Unions would retain the same functions as under private enterprise but public ownership meant they would have obligations to the wider community. Unions would be jointly responsible with management for ensuring the efficient running of the industry and maintaining discipline amongst the workforce. Collective bargaining would expand, but as the profit motive would no longer predominate there would be no exploitation and so a new co-operative system of industrial relations would emerge.[59]

With the adoption of *Coal: The Labour Plan* the debate subsided. The re-election of the Conservative National Government in 1935 meant public ownership was not on the political agenda and until there was a realistic prospect of a majority Labour government there was no point in taking the matter further. Nationalisation was raised at the 1941 MFGB Conference as part of the wider debate over the industry's production problems but the Executive recognised that the Coalition would not nationalise the coal industry and so there was no point in pursuing it further.[60] Thinking on the post-war structure of the coal industry was stimulated by the approaching end of the war in Europe. The TUC's 1944 *Interim Report on Post-War Reconstruction*, for example, proposed that public ownership be based on the 'full participation by workpeople and their representatives in the affairs of the industry.' Public ownership would promote the common good, improve living standards and democratise industry, and the *Interim Report* identified the public corporation as proposed in the TUC's 1932 document as the preferred organisational model, explicitly citing Morrison's London Passenger Transport Board as an exemplar.[61]

The Governmental Response

In May 1945 Gwilym Lloyd George, the Minister of Fuel and Power in the Coalition government, told the War Cabinet there was no dispute over 'the importance of re-establishing the British Coal-mining industry.' Citing the Reid Report, Lloyd George reeled off a catalogue of familiar shortcomings: 'Recruitment is unsatisfactory, both in numbers and in quality; labour relations remain poor; output per manshift is far too low and the cost of production uneconomically high.' The government accepted Reid's recommendation for a National Authority with legal powers to concentrate and plan production and to work the country's coal reserves efficiently. This decision 'would in no way prejudice the free choice of the electorate in the question of ownership' but would 'show that all parties were resolved to restore as soon as possible the efficiency of the British coal-mining industry.'[62] On 23 May the Coalition ended and the General Election was scheduled for 5 July. In the House of Commons on 29 May Gwilym Lloyd George announced that Churchill's Caretaker Government intended to use the Reid Report as the basis for post-war coal policy and that the industry would remain in private ownership. Although the Government accepted the need for compulsory reorganisation, the day-to-day operation of the industry would remain the responsibility of those who produced the coal.[63] This was the result of pressure from the Chairman of the Conservative Party, Ralph Assheton, who had, in turn, been pressed by party activists and MPs to keep the coal industry in private hands. Naturally, this provoked a furious response from the Labour benches who saw this as Churchill doing the MAGB's bidding, arguing that the policy owed more to Foot than Reid and that the policy was merely a continuation of the discredited solutions of the 1930s. The MAGB Central Committee had (on 14 June) accepted Lloyd George's statement in full as the basis for its post-war policy. The commitment to compulsory reorganisation led in July 1945 to Foot modifying his original proposals in order to give the proposed Central Authority powers to compel amalgamations and reorganisation.[64]

The NUM had welcomed the Reid Report and endorsed its conclusion that technical changes could not be carried out by the industry as then constituted. Reorganisation would not happen unless there was a powerful central body willing and able to override sectional interests. Logically, the NUM argued, this pointed to public ownership and it recalled that both the Sankey and Samuel Commissions (1919 and 1926) had advanced analyses similar to Reid's but subsequently little had happened. The NUM regretted 'the wasted years; if the technical and reorganisation proposals we put forward then had been put into operation the industry would not have been in the plight it is today.'[65] The TUC/Labour Party/NUM committee preparing the movement's policy was chaired by Morrison, and Emmanuel Shinwell, soon to be Minister of Fuel and Power, was a member. It met five times in the spring of 1945 and its deliberations were well advanced by the time the General Election took place and were completed soon after the result was declared (26 July). Labour's manifesto, *Let Us Face the Future*, gave an unequivocal pledge that the industry would be nationalised:

> For a quarter of a century the coal industry, producing Britain's most precious national raw material, has been floundering chaotically under the ownership of many hundreds of independent companies. Amalgamation under public ownership will bring great economies in operation and make it possible to modernise production methods and to raise safety standards in every colliery in the country. Public ownership of gas and electricity undertakings will lower charges, prevent competitive waste, open the way for co-ordinated research and development, and lead to the reforming of uneconomic areas of distribution.

When it became clear Labour was going to form a government the NUM Executive and Shinwell agreed not to publish the joint report but Shinwell did agree to regard the proposals as 'the views of the movement' and take them into account when preparing the nationalisation bill.[66]

The Conservative manifesto was far more detailed than Labour's in its proposals for the industry. These followed Lloyd George's statement closely and endorsed Reid's analysis of the industry's ills but it ruled out public ownership:

> The industrial activities of this country are principally founded on coal. Adequate supplies, as cheap as possible, must be available for our homes, for our factories and for export....Wartime measures are not suited to peacetime conditions. A new, practical start is needed. The position cannot be remedied by mere change of ownership of the collieries. That offers no solution:

A central authority, appointed by the Minister of Fuel and subject to his general direction, would be created to ensure that reorganisation took place. Reorganisation would centre upon the development and efficient conduct of coal mining according to the best modern practice and 'In so far as grouping or amalgamating collieries is necessary for this object, it will be carried through, voluntarily if possible, but otherwise by compulsion.' However, 'we do not propose amalgamation for amalgamation's sake.' Reorganisation was to be the responsibility of the industry with the central authority satisfying itself that the plans conformed to national requirements but the Central Authority would have 'powers of enforcement in reserve.' This policy would:

> preserve the incentives of free enterprise and safeguard the industry from the dead hand of State ownership or political interference in day-to-day management. It will also provide the necessary sanctions for making sure that the essential improvements recommended in the Reid Report are carried through.

With the exception of the issue of who was to own the industry, both the Labour and Conservative manifestos drew their inspiration for their policies from the Reid Report. Its emphasis on technocratic renewal did, however, mesh with the Morrisonian model of the public corporation.

The history of the mines' nationalisation is coloured by Shinwell's comment in his memoirs that when he arrived at the Ministry of Fuel and Power all he found were a few pamphlets on nationalisation, and that despite talking about public ownership for decades the Labour Movement had no clear proposals.[67] This is disingenuous. The basic principles upon which the industry was to be nationalised were thought out and were available, but given the sheer complexity of nationalising coal it is not surprising that the movement (including the miners) fought shy of formulating detailed plans until actually in office; the details of coal nationalisation would inevitably absorb a massive amount of political energy. Morrison's role as public ownership overlord through his office as Lord President of the Council and chair of the Cabinet Socialisation of Industries Committee ensured that 'the Morrisonian model' (with suitable variations) underpinned the Attlee government's nationalisation programme.[68]

Without coal, post-war recovery could not happen, but the industry was riddled by an accumulation of problems and discontents which threatened to spill out into the wider political economy. Controlling this discontent was a major objective of public ownership. Central to the management of discontent in the coal industry were two principles. The first was that industrial relations and collective bargaining were the joint responsibility of management and union and not of any third party (including government); and second, that management and union had an obligation to act in accordance with the national interest. The NCB's structure and ethos, with its emphasis on self-government and arm's length ministerial involvement, could easily embrace these two principles, which had also lain at the heart of the development of the state's relationship with organised labour over the previous fifty years. The coal industry's mechanisms of conflict management, the consultation and conciliation systems, were a reflection and institutionalisation of practice concerning traditional state-union relations rather than a radical departure. Nationalisation's radicalism stemmed not from any innovation in industrial relations but in the expropriation of the private owners, and despite the rhetoric the basic distribution of power between management and labour remained unchanged.

Notes

[1] D. Howell, "All or Nowt": The Politics of the MFGB', in A. Campbell, N. Fishman and D. Howell eds, *Miners, Unions, and Politics 1910-1946*, Scolar Press 1996, pp. 36-7.
[2] For the politics of the MFGB in the late 1930s see, A.J. Taylor, "Maximum Benefit for Minimum Sacrifice": The Miners' Wage Campaign', in *Historical Studies in Industrial Relations* vol. 1 no. 2 1996, pp. 65-92, and N. Fishman, *The British Communist Party and the Trade Unions 1933-45*, Scolar Press 1995, pp. 164-99.
[3] H. Francis, 'Learning from Bitter Experience: the Making of the NUM' in, Campbell, Fishman and Howell eds, *Miners, Unions and Politics*, pp. 253-72.
[4] MFGB, *Annual Conference Report*, July 1942, pp. 41-2. The debate is at pp. 40-63. Only Nottinghamshire opposed the Executive's initiative.
[5] MFGB, *Report of a Special Conference*, 16-18 August 1944, pp. 91-2.
[6] MFGB, *Report of the National Executive*, June 1943, pp. 213-14.
[7] *Report of a Special Conference*, August 1944, p. 81.
[8] *Report of a Special Conference*, August 1944, p. 86.
[9] *Report of a Special Conference*, August 1944, pp. 62-3.
[10] *Report of a Special Conference*, August 1944, p. 19.
[11] *Report of a Special Conference*, August 1944, pp. 20-21.
[12] *Report of a Special Conference*, August 1944, p. 63.
[13] *Report of a Special Conference*, August 1944, pp. 124-26.
[14] P. Gibbon, 'Analysing the British miners' strike of 1984-5', *Economy and Society* vol. 17, 1988, pp. 152-4 for a similar distinction.
[15] Compositing was the process whereby aspects of a common issue or several related issues were amalgamated into a single resolution to reduce the time spent discussing an issue. In the NUM the initiative for compositing came usually from the Conference Business Committee and NEC.
[16] *Report of a Special Conference* p. 39.
[17] Benyon, 'The Making of the NUM', pp. 268-9.
[18] *Report of a Special Conference*, p. 21.
[19] W.H.B. Court, *Coal*, HMSO/Longman Green & Co. 1951, p. 391.
[20] R. Page Arnot, *The Miners. One Union, One Industry. A History of the National Union of Mineworkers 1939-1946*, George Allen and Unwin 1979, p. 81.

21 C. Barnett, *The Audit of War. The Illusion and Reality of Britain as a Great Nation*, Macmillan 1986, pp. 63-86 documents more briefly than Court the depressing condition of the wartime coal industry.
22 Ministry of Fuel and Power, *Coal Mining. Report of the Technical Advisory Committee* Cmd.6610, March 1945, para 1. Hereafter, *Reid Report*.
23 *Reid Report*, para 2.
24 *Reid Report*, para 193.
25 *Reid Report*, para 205.
26 *Reid Report*, para 207.
27 *Reid Report*, paras 675-6.
28 *Reid Report*, paras 683-8.
29 *Reid Report*, paras 692-6.
30 *Reid Report*, para 749. Emphasis added.
31 *Reid Report*, para 760.
32 W.A. Lee, *30 Years in Coal. A Review of the Coal Mining Industry under Private Enterprise*, Mining Association of Great Britain 1954, pp. 165-8.
33 This was Robert Foot, formerly Director General of the BBC (1941-1944) and chairman of the Gas, Light and Coal Company Ltd (1929-1941). As a condition of his employment Foot insisted that he be given a free hand.
34 R. Foot, *A Plan For Coal. A Report to the Colliery Owners*, Mining Association of Great Britain, January 1945, para. 1-2, p. 1. Hereafter, *Plan for Coal*.
35 *Plan for Coal*, para 23.
36 *Plan for Coal*, para 81.
37 *Plan for Coal*, para 129.
38 *Plan for Coal*, para 132.
39 *Plan for Coal*, para 137.
40 Lee, *30 Years in Coal*, p. 196.
41 The Owners' Plan for Coal, *The Economist*, 27 January 1945, p. 103.
42 *The Economist*, 27 January 1945, p. 103.
43 NUM, *National Executive Committee minutes*, 7 February 1945. Hereafter *NUM (EC)*.
44 Appendix. The Coal Plan of Mr. Robert Foot. A Preliminary Memorandum of the National Union of Mineworkers, *NUM(EC)*, 7 February 1945. This was circulated under the signature of Ebby Edwards, the General Secretary, but was written by Moffat, Hall, and Horner.
45 *NUM (EC)*, 22 February 1945.
46 *The Miners' Case. An Answer to the Foot Plan*, National Union of Mineworkers, 1 March 1945.
47 *NUM(EC)*, 8 March 1945.
48 This section is derived from A.J. Taylor, 'The Miners and Nationalisation' in, *International Review of Social History*, vol. 28 no. 2 1983, pp. 176-99.
49 Quoted in D. Coates, *The Labour Party and the Struggle for Socialism*, Cambridge University Press 1975, p. 19, and Report of a Meeting between the MFGB Executive and Officials of the Parliamentary Labour Party, 26 March 1929, in MFGB, *Executive Committee minutes*, 12 April 1929.
50 Lord Morrison of Lambeth, *Herbert Morrison. An Autobiography*, G. Allen & Unwin 1960, p.119.
51 Labour Party, *Report of the Annual Conference 1932*, pp. 212-14.
52 H. Morrison, *Socialisation and Transport*, Constable 1933, pp. 210-12, and pp. 233-39.
53 Trades Union Congress, *Report of the Annual Congress 1932*, pp. 206-19 and pp. 394-96.

54 Trades Union Congress, *Report of the Annual Congress 1933*, p. 84 and p. 210.
55 Labour Party, *Report of the Annual Conference 1933*, p. 265 and MFGB, *Executive Committee minutes*, 11 May 1933.
56 MFGB, *Report of the Executive Committee 1933*, pp. 26-27.
57 E. Edwards to W. Citrine, 22 December 1932 in MFGB, *Report of the Executive Committee 1933*, p. 28.
58 MFGB, *Executive Committee minutes*, 15 May, and 20 May 1936.
59 The TUC/Labour Party policy was published in 1936 as *Coal: The Labour Plan*.
60 MFGB, *Executive Committee minutes*, 20 November 1941.
61 *Interim Report on Post-War Reconstruction*, Trades Union Congress 1944, paras. 37-38, and *NUM (EC)*, 12 April 1945.
62 W.P. (45) 308 16 May 1945. Future of British Coal Mining. Memorandum by the Minister of Fuel and Power. *CAB 66/65*.
63 5s *H.C. Debs* 411, 29 May 1945, cols. 87-8.
64 Lee, *Thirty Years in Coal*, pp. 199-200.
65 NUM, *Press Statement on the Reid Report*, 12 April 1945.
66 *NUM(EC)*, 17 August, and 20 September 1945.
67 E. Shinwell, *Conflict Without Malice*, Odhams 1955, p. 172. Slowe describes Shinwell's claim as 'extraordinary' and 'a lie designed to make his achievements seem greater than they were. There may not have been an actual blueprint, but there were a lot of useful and detailed guidelines.' P. Slowe, *Manny Shinwell: An Authorised Biography*, Pluto Press 1993, p.208.
68 P. Hennessy, *Never Again. Britain 1945-1951*, Pantheon Books 1993, pp. 198-201 disputes the existence of the 'Morrisonian Model' applied mechanically to the nationalised industries. He is both right and wrong; there was a model but it was applied with reference to the situation in each industry.

Chapter 2

Inclusion or Integration?

Introduction

It is now difficult to recall or appreciate the centrality of coal to Britain's political economy as it entered the post-war era. British coal was fundamental to the economic recovery of both Britain and Western Europe, and the 700,000 mineworkers and the NUM were a significant block of political and economic power whose behaviour had enormous implications for post-war British politics. The first part of the chapter examines the distribution of power in the coal industry and in particular, the NUM's exclusion from decision making on core issues. The next two parts examine the NUM's relationship with the state with respect to the problems of coal output and the importance of mineworkers' wages which impacted on the price of coal, general wage levels and inflation. Finally, the chapter considers the consequences of the Cold War on NUM politics and union-government relations. Permeating these issues is the question of the extent to which involvement with the state should determine NUM policy. In these early years many of the attitudes and problems which were to dominate the NUM's politics and relationship with the state were established.

Old Wine in New Bottles?

Despite the veneer of socialist rhetoric and expectations the public corporation as it emerged and developed in the 1940s was a technocratic and conservative phenomenon compatible with capitalist political economy. Nevertheless, the nationalised industries were an important political development. First, they were part of the state's response to the rise of class politics by incorporating organised labour itself into the management of the economic and political consequences of labour's rise. They were therefore a species of class compromise and accommodation and this was particularly important in the case of the NUM and the coal industry. Second, nationalised industries such as the NCB were compatible with an overwhelmingly privately owned economy. The industries nationalised by the Labour government were rundown and unprofitable but essential to the success of private manufacturing. If private enterprise could not sustain the coal industry the state would have to shoulder the burden, although this was complicated by the ideological baggage associated with nationalisation in the labour movement. Publicly owned industries posed few serious doctrinal problems for the Conservative Party or private industry. Third, although part of the liberal democratic state, publicly owned industries were once removed from the political system's representative institutions, and as technocratic-managerial para-state institutions they were part of its administrative bureaucratic structure. Fourth, the ideology and ethos of the traditional managerial and business élites predominated in these industries. Senior NCB management was socially and culturally closer to key political decision makers and administrators than union officers and became more so as the state-industry relationship evolved and orthodox business ethics and accountancy practices determined the governance of these industries.

The NUM was well represented in the formal liberal democratic political system but hardly at all in the state. The NUM had a presence on the Labour Party's NEC, the TUC General Council and was one of the 'Big Six' unions in the labour movement in the 1940s and 1950s. Even when the NUM began to shrink it remained a powerful symbol in the labour movement and a major contributor to Labour Party funds. On the other hand, the NUM's presence in the administrative state was weak. Irrespective of their party, ministers were determined that the Ministry of Fuel and Power would not be responsible for running the industry or for its industrial relations. This would undermine the NCB's authority and responsibility for the industry by weakening the arm's length principle and would focus NUM attention on government rather than the NCB. The ministerial-board relationship was a sectoral corporatism in which overall policy was determined centrally but implemented by the NUM and NCB acting in concert, an arrangement acceptable to both Labour and Conservative governments. These weak linkages were reinforced by government's refusal to lay down a fuel policy and the lack of technical expertise in the Ministry of Fuel and Power to challenge that of the NCB.

Government control of the coal industry had focused political attention of the MFGB and NUM on government as the arbiter of policy and ministers believed strongly 'It is essential to break this habit.'[1] Ministers recognised that 'since the accounts and reports...would be laid before Parliament, and the general responsibility of Ministers was to Parliament' there would be debates on the nationalised industries but core decision making on investment, planning, prices and industrial restructuring was insulated from parliamentary control. The government's majority made scrutiny purely formal and Dalton agreed with Gaitskell that 'if the Public Accounts Committee showed any signs of over-meticulous enquiry, it could be warned to moderate its ways.'[2] Despite its importance in the post-war political economy the NUM was isolated from the key bureaucracies. Meetings with ministers occurred only when there was a crisis in the industry and in these the NUM was invariably on the defensive. Normally relations were mediated through the NCB or bodies such as the Miners' Parliamentary Group. By December 1947 all work relating to coal production had been transferred from the Ministry of Fuel and Power to the NCB and despite its 'lead' role in industrial relations the Ministry of Labour played no role in the industry as the coal industry had its own system of conciliation. The NCB was a tripartite structure in that ministers, management and unions were involved in the industry's processes but there was no tripartite forum where all three met together. The strongest relationship was the informal minister/board relationship; the management-union relationship was highly institutionalised but was not part of the industry's normal pattern of bureaucratic politics, so for all practical purposes the minister-union relationship was non-existent. The publicly owned coal industry was a *bifurcated tripartism* in which core decisions on investment, planning, prices and industrial restructuring remained under managerial control within the administrative state with the unions confined to consultation and the legitimisation and implementation of managerial decisions. The NUM enjoyed status not power, its participation in the coal industry was severely circumscribed and the industry's processes and structures not only failed to resolve management-union conflict but actually exacerbated conflict within union. This was the result of the failure to transform the culture of conflict in the coal industry and to demonstrate unequivocally its superiority to private enterprise. The NCB remained a politically contested organisation but was subjected to few serious ideological assaults. When structural change took place it was incremental, change in the light of experience or as a result of the industry's decline rather than any fundamental reappraisal. This had to wait until the 1980s.

Nationalisation transferred ownership from the private to the public sector but ownership is of secondary importance when compared to actual control. The publicly owned coal industry was part of wider shift to welfare capitalism and the NCB's power structure was not markedly different from any other large-scale enterprise. Between 1947 and 1953 NUM/NCB relations developed into an inter-organisational relationship between two parallel but inter-connected bureaucracies. In this relationship 'there remains this undeniable fact that the Coal Board is an employer and we are a Trade Union... They buy our labour power and we sell it, and of course we sell it dear. They, as the employer must do, endeavour to buy it at a lower price than at which we are ready to sell. That conflict is there. We must try to handle it sensibly – not in the old way, but in an intelligent way.'[3]

The 1945 NUM Conference was held in the last week of June. In his Presidential Address Will Lawther drew an explicit comparison with the state of the industry in 1918 and ominously, compared the end of the 1944 National Wages Agreement due in 1948 to the end of government control in 1921 which had culminated in 1926. A re-elected Conservative government threatened a 'drift back to the economics of the jungle' and this was not acceptable to the NUM, only a Labour government committed to nationalisation offered the mineworkers and the nation security. Lawther insisted nationalisation would, of itself, generate a cultural revolution amongst the workforce for 'once the legislation is passed, then we shall have the chance to tackle the job we have longed for – to give the nation a mining industry of which it can be proud, our industry producing plentiful supplies of coal at reasonable prices under conditions that provide a healthy and happy life for those who have to work in it worthy of their arduous calling.'[4] Whether the industry was capable of meeting these expectations was unclear and depended to no small degree on the structures in which working relationships were located. By the end of October 1945 Shinwell had decided on the industry's basic structures but the nationalisation bill dealt with only the most essential matters such as the powers and composition of the Board, leaving much of the detail to be decided later. There would be a regional structure and there would be a distinction between financial and administrative autonomy in that the industry's finances would be centrally controlled to permit the cross-subsidisation of coalfields. Prices would also be determined nationally. The speed of nationalisation and the vagueness of the Coal Industry Nationalisation Act (CINA) gave the NCB considerable freedom to determine its *modus operandi*.[5]

After the bill's publication the NUM's representatives on the Joint National Negotiating Committee (JNNC) requested clarification on several points. The most important were concerned with the Minister's power to make regulations and in particular they felt that 'the provision for establishing consultative machinery...should not be left to Ministerial regulations, but should be provided for in the Bill itself.' Second, the NUM argued strongly that it 'should be consulted in the appointment of certain members and given an opportunity of making observations on the membership of the Board as a whole.' Third, the NUM did not want centrally appointed Regional Directors responsible to the NCB but Regional Boards with local responsibility. Fourth, there was no mention in the bill of the amount of compensation to be paid to the industry's existing owners and the NUM was anxious to ensure this was as limited as possible. Finally, the power of the minister to give 'directions of a general nature' to the Board was not clearly defined.[6]

The meeting with Shinwell which discussed these queries was, in effect, concerned with the distribution of power in the publicly owned coal industry. Ministerial power to make regulations, Shinwell argued, was essential because of the NCB's complexity and to include everything in the bill would delay its passage. Regulations were, in any case, subject to parliamentary approval and would have to conform to the principles of the Act. On consultation Shinwell insisted 'it was quite inconceivable that the Board would not enter into consultation with the representatives of the workpeople' and consultation and collective bargaining was implied in the Bill. On appointments the government 'was not prepared to consult the Union on the Board as a whole...the persons so appointed would be representatives of the Nation and not any particular section of the community.' Shinwell agreed to consult the TUC on the appointment of one Board member and the NUM would be asked to suggest one person 'who had had experience in the organisation of workers', and there would be a 'labour' representative in each of the regions. Pit managers would be expected to consult NUM branches but the 'day-to-day management of the pit...must always remain the responsibility of the manager.' The details of this consultation system would be worked out between the NUM and the NCB.[7]

The classic study of corporate power structures defined effective control as the ability to determine the composition of the directing board which, in coal's case, lay with the minister.[8] Studies of corporate governance divide control into strategic and operational, with the former relating to determining the corporation's operational parameters (the minister) and the latter, the implementation of corporate strategy and conducting day-to-day operations (the NCB). This left no legally sanctioned place for the NUM within the structure of control but the NUM could not be left out of the control equation so its role was expressed not by legal but by social, political, economic and even ideological, means. This was reflected in the consultation and conciliation procedures. Originally the nationalisation bill contained no legal requirement that the NCB establish any consultative machinery. Pressure from the NUM and the Conservative Opposition's use of this omission to attack the credibility of nationalisation forced Shinwell to agree to a clause requiring the NCB to establish consultation and conciliation machinery.[9] The NEC's view was that the Bill was generally acceptable but there remained a significant undercurrent of disquiet concerning many issues, notably consultation, collective bargaining, and appointments. Lawther's assessment of the NCB was ruthlessly pragmatic as 'whatever may be the composition of the Board, whatever titles they may be adorned with, whatever may be their outlook and whatever they may say to us, they will never get within miles of the hard-faced men we have met in the past.'[10]

Nationalisation was an epochal event. A new era was dawning as 'within a few months the word "coal-owner" will pass out of our language. We enter this new era in which both the coal we win from Mother Earth and the mines become the property of the nation. Nationalization of the mines is the realisation of our hopes and aspirations.' A Durham delegate believed public ownership would abolish the class war in the pits, but warned that 'there is a need for propaganda and a facing up to the true position of our future responsibilities. All frustration and bitterness must be swept away.' He concluded by calling for 'a new industrial morality, to secure friendly discipline within the industry.'[11] Horner called for an end to the culture of conflict because even though the NCB was an employer, it was a different kind of employer and 'We must take a long view, because the advancement of our members' interests lies not in sneaking victories from coalowners, but in establishing a firm and highly productive industry...That is a different role for our members...the day of agitation...is nearly finished. The task in future is to use the best means we know to fight Mother Nature and to drag the coal out of her bosom. The future that lies before us is tremendous' and Horner concluded 'The pits are ours. We can say what can be done with them.'[12]

Vesting Day on 1 January 1947 was not marked by the NUM. The Ministry of Fuel and Power had contemplated holding mass meetings in large cities adjacent to the coalfields but the NEC declared they would serve 'no good purpose' and would disrupt production. At some pits there were ceremonies (as at Horden where management and union buried a literal and metaphorical hatchet) and speeches as the NCB's flag was hoisted. Whilst the coal-owners' departure was a matter of no regret and public ownership was seen positively many mineworkers going on shift on 2 January could be forgiven for noticing that the same faces remained in charge of the pits. Vesting Day had a brief positive effect on attitudes and output but this soon faded. This, a contemporary study argued, was because nationalisation was such a powerful ideal no reality (especially in the conditions of 1946-47) could hope to match the ideal. Complaints soon began to flow about too many bureaucrats and there being no real change in management-workforce relations, except that it was now more difficult to resolve conflict without the involvement of a higher authority. As one miner put it: 'We don't know who the gaffers are now.'[13]

The NCB was a large, vertically integrated corporation managed by technocrats under the arm's length tutelage of a minister. The minister was empowered to appoint the national Board under Section 2 of the 1946 Act; Section 3 laid down the minister could, after consulting the Board, issue general directions on matters that appeared to the minister to affect the national interest. There was no formal organisational link between the minister and NCB, so consultation took the form of informal meetings. The power to issue general directions was never used but the act enabled ministers to determine the NCB's operational parameters whilst leaving implementation to the Board allowing them to disclaim any responsibility for the Board's actions or policies. This was a fiction, but a politically important fiction. The minister had the ultimate power to overrule the NCB but would never do so formally as this could provoke mass resignations or result in the ministry assuming responsibility for coal. Ministers retained strategic control over the coal industry through largely indirect means, and the NCB's growing dependence on the Treasury was also an important factor.

In July 1947 Herbert Morrison, the Lord President and chairman of the Socialisation of Industries committee, circulated *Taking Stock*, one of the founding documents of the post-war mixed economy. Morrison believed it was 'generally agreed' there should be as few ministerial directions to the public corporations as possible but they must 'act on lines settled from time to time with the approval of the minister.' Morrison saw informal relations as the means 'to ensure that Ministers are able to influence the Boards' and the power of appointment was particularly important.[14] In his response to *Taking Stock* Hugh Gaitskell (Shinwell's successor) pointed out that the complexity of transferring coal from the private sector meant there had been 'continuous informal contact, between his ministry and the NCB and he had had no reason to issue general directions. CINA gave operational control to the NCB whose monopoly of technical expertise could not be challenged by ministers because of the shortage of mining engineers who were better employed mining coal than scrutinising the NCB. Ministerial control of production had ceased on 1 January 1947 and the ministerial headquarters coal staff had been broken up, so informal relations between the NCB chairman and Vice-Chairman with the Minister and Permanent Secretary were the main linkage. Gaitskell emphasised that Lord Hyndley, the NCB chairman, had been 'careful to seek informal guidance on matters on national or public importance' and commented that he (Gaitskell) did not want to see relations evolve beyond this point.[15] Privately Gaitskell confessed that establishing a *modus vivendi* with the Coal Board was not easy. He was not unsympathetic to the views of some ministers (for example, Aneurin Bevan) for greater ministerial control but 'it is no easy job to try and establish just the necessary degree of control without going too far. Also it is irritating not to be able to keep them on the right lines all the time.'[16]

A central purpose of public ownership was to remove coal from the direct orbit of government. In one of Gaitskell's first meetings with the NUM Executive he urged the speedy disengagement of the NUM from the Ministry because the new consultative and conciliation procedures meant that 'the functions concerned with production and labour, as had been the responsibility of the Ministry during the period of control, had now passed to the [NCB].' Coal's importance (and the NUM's role in the Labour Party) ensured the continuation of a political linkage which meant 'he would always be willing to discuss any matter relating to the industry' with the NUM and as minister he had 'to take a close interest in coal production.'[17] The NUM accepted disengagement, especially after the incomes policy of 1948-50 which threatened to make the Minister a direct bargaining partner. The NUM agreed 'it had always been considered inadvisable that the Minister should have any authority to intervene on a question of wages or conditions of employment.'[18]

Decisions taken by the NCB had economic and political implications which the government could not ignore. In the summer of 1948, for example, the NUM and NCB were negotiating a supplementary pension and redundancy scheme designed to ease the restructuring of the industry. Not only was this costly but it was feared it might affect output as labour left the pits. Gaitskell's dilemma was not *whether* to interfere but *when*, for 'The present procedure is that we do not interfere until the negotiation is complete.' To avoid the charge of interference 'we have had to give private advice to the NCB, which the NCB were not always disposed to accept.' In any case the NCB was often reluctant to act as a 'screen' for the government, fearing this might affect relations with the NUM Gaitskell wondered if it would be easier to have earlier formal contact, but what if the NCB still rejected his advice? Gaitskell concluded 'the next step is to muscle in on discussions quite openly; even if we can't persuade the miners and the board'.[19] There was no easy solution: 'It is really very unsatisfactory having to deal with the Boards as though they were independent authorities with no special obligations to the government. Yet if they choose that is the line they can take, and without a major public row there is very little the Minister can do about it.'[20] Radical change so soon after Vesting Day would be interpreted as an admission of failure, so as the government was locked into the public corporation model every effort had to be made to make it work effectively.

Further experience with public ownership confirmed Gaitskell's conviction ministers should not issue directives and that his powers of appointment and dismissal were limited. Conservative attacks meant the industry had to be kept at arm's length from the minister and the minister's relationship with the Board were structured around three principles: there should be no duplication of staff by the ministry and the Board; the Board should be the source of ministerial advice on the coal industry; and management should not be 'mothered.' Gaitskell 'did not believe a Minister should, at least in any formal sense, interfere with the Boards so far as they are merely endeavouring to carry out their statutory obligations. He must do his best to pick the right people to be members of the Boards and then give them a pretty free hand to carry out the duties entrusted by Parliament to them.'[21] Within this structure the most consistent contact ('Week in, week out') was between civil servants and Board officers. These contacts could 'be carried much further' and even though a minister could not direct the Board 'he may find it perfectly easy and natural to discuss privately with the Chairman all sorts of problems.' These contacts were regularised into weekly meeting of between two and three hours during which 'we talk over a whole range of questions, some particular, some general, some raised by the Chairman and some raised by myself.' With such a range of informal contact 'there is, or should be, no reason why discussion should not cover virtually any part of the Board's activities.'[22] There was no similar contact with the NUM.

The Problem of Coal Production

The demand for coal produced a powerful tension between the NUM's immediate concerns and modernisation, and it became a common complaint in the NUM that short-termism prevented strategic thinking. At the 1947 NUM Conference Jim Hammond (Lancashire) complained 'they were living from hand-to-mouth...settling a dispute here, settling a dispute there, wrestling with a price list here'; Will Arthur (South Wales) took the opposite tack arguing the immediate problems of the industry could not be addressed because the NCB were 'up to their ears in an examination of long-term planning.'[23] The political pressure for maximum production was irresistible. Lawther compared the political importance of fuel supplies with wartime weapons production when 'the most ruthless measures were quite rightly adopted, and it did not matter whose corns were trodden on' and the Labour government was under no illusions about the coal problem.[24] Shinwell and the NUM were convinced 'that once the industry was nationalised, miners would work harder and more willingly' because the coal-owners had gone so the miners would receive a better deal and because *their* government and *their* union exhorted them to greater effort.[25] This rested on the assumption that public ownership *per se* would generate a cultural change, a proposition which found little support in, for example, Reid's analysis or that of officials at the Ministry of Fuel and Power. Shinwell 'did not expect the benefits of nationalization to affect the men immediately' but substantial benefits would be forthcoming. For their part the NEC declared they were 'not unmindful of their responsibilities' and agreed to do all they could to boost output.[26]

In October 1945 Horner had outlined the NUM's strategic role. 'Under all circumstances, whether under private ownership or national ownership,' Horner argued, 'we exist for one purpose and one purpose only, and that is to safeguard and improve the working and living conditions of our members.'[27] The issue was how public ownership would change this role. Nationalisation had to deliver the coal needed to ensure the survival of the Labour government and Horner argued the actions of the NEC should be judged against this criterion. With the coal-owners removed there should now be a 'new morality in regard to production' and in promoting this new attitude 'It is not a question of bamboozling the membership: it is a question of telling the truth and putting things in their correct perspective, relating present action to future possibilities.'[28] Jim Hammond introduced a note of caution, 'we have to face our own rank and file and I may say that I find today that there is sometimes a greater struggle with our own rank and file than we had with the employer. We have to realise – and it is quite logical – that there is natural tendency for our rank and file, having voted a Labour government in, to think that rabbits will jump out of the hat at once.'[29] These speeches encapsulated the conflicts inherent in the politics of production. Zweig reports the reaction of a mineworker which demonstrates the passions aroused: 'No one can bind me to do more work,' shouted a Kentish miner at me in a frantic voice. 'We all resent that. My representatives are here to defend me, but not to knock me down or run up a bill for me. Oh no. If have to pay the bill, I have to draw it for myself'.[30] The government had called for an extra 8m tons of coal for the winter of 1945-46 but output only increased by 2m tons. Resolving these tensions was the most difficult problem facing the NUM.

The NUM's *Charter of Demands*, formulated in December 1945, was designed to contrast the new and old orders, improve recruitment and thereby boost output. Shinwell asked the NEC what incentives were needed to boost output and the NEC outlined a programme which focused on the industry's modernisation, sought a radical re-think of the wage structure, a wide range of social benefits, a seven hour day underground, a five day week, a guaranteed weekly wage with miners at the top of the wages league, and better housing. The country's 'total dependence' on coal, the NUM argued, meant the government ought to implement these measures according to a timetable and, as coal was a national responsibility, the mining communities were unwilling to meet their obligations unless the nation played its part.[31] The Charter posed a serious political problem for the government. Shinwell wrote to Lawther with the government's response. Nationalisation, he wrote, would eventually bring about the changes sought by the NUM but these changes would have to come about through negotiations between the NUM and NCB, not with the government. On a range of issues, the five day week, paid holidays, wages, and compulsory trade union membership, the government was determined not to increase costs or do anything which might reduce output; this reply so worried the NEC they sought a meeting with Shinwell to secure 'satisfactory assurances.'[32]

The NUM believed strongly that the coal would be forthcoming if concessions were made, whereas the government believed concessions should follow increased output. The government was reluctant to institute the five day week as this would cost 15m tons of coal per year. Three key issues – holidays, wages, and compulsory union membership – Shinwell insisted were matters for negotiation between the NUM and NCB and neither he nor the Cabinet would object if the NUM and NCB reached an agreement. The NEC noted Shinwell's willingness to improve conditions in the industry but that in return 'he must be assured that the output necessary for the country's well-being would be forthcoming.'[33] The Labour Government was most reluctant to be drawn into the industry's internal bargaining and refused to act as a surrogate employer, and regarded it as a matter of the gravest political urgency that as many issues as possible be transferred speedily to the NUM and NCB. The NEC agreed the conciliation machinery should be based on the 1943 agreement and that it should present the Miners' Charter to ministers with priority being given to the five day week. The negotiation of the Five Day Week Agreement (FDWA) became the first test of the new industrial relations system in the coal industry as the NUM refused to submit this to arbitration by the National Reference Tribunal (NRT).[34] The main problem was the estimate that the FDWA would cost 15m tons of coal a year but it had become symbolic of the difference between the private and publicly owned industry. At least one delegate feared that the presence of a Labour government was affecting the NUM's willingness to defend its members: 'if we were under private enterprise we should be fighting much harder for this'.[35]

The NCB was willing to discuss the five day week but warned the NUM it had to take account of its consequences for coal output and prices (the government's concerns) and could not commit itself to an early announcement. The NUM sought an announcement on Vesting Day that the five day week would be implemented on 1 May but conceded that if the NCB could not do so negotiations should begin immediately, which they did. The NUM argued government had accepted the principle of the five day week on condition output was not damaged (the union's view was that output would increase) but 'the change to a five-day working week must be brought about irrespective of any question of cost.' The five day week was approved at the first meeting of the Joint National Negotiating Committee on 17 December 1946 and in response to Lord Hyndley's call for co-operation and mutual trust Lawther responded that 'The NUM were aware they were now dealing with men who appreciated the needs of the Industry.'[36] The Five Day Week Agreement came into force on 1 March 1947 despite concerns on the Board and in Cabinet about the possible loss of the average output of 350,000 tons of coal mined on Saturdays. Mineworkers would be paid an extra bonus shift if they worked five full shifts and full shift working coupled with the morale boost given by the five day week would make up the lost output.

The importance attached by the NEC to the five day week as a symbol of the new order can be seen in Horner's stress on the willingness of the NCB to negotiate which meant the NUM had a responsibility to make up the lost output. The political significance of the agreement was that it had received Cabinet approval in the expectation that the 350,000 tons of coal produced every Saturday would be made up in normal shift time. The NCB and government were taking a gamble wondering 'what will happen to the country if we lose that 14m tons? Well, I think it will bring down the Labour government, because industry could not function. It would not be a matter of putting out the lights for a hour a day.' The five day week was 'the greatest single reform ever to come into this industry', a reform which had not been forced out of the NCB but had been secured because the Board (and government) had been persuaded by the merits of the NUM's case. Successful implementation would not only help Labour's cause but show that public ownership was far more effective than private ownership in delivering the coal the country needed.[37] This was the background to the fuel crisis in the winter of 1947.

The mines had been transferred to public ownership on 1 January 1947 and on 23 February the country's power generation and transmission systems virtually collapsed under the impact of the severest winter in living memory. The Ministerial Coal Committee had adjourned at the end of March 1946 having convinced itself that as a fuel crisis on the lines of 1944-45 had been averted in 1945-46, they could look forward optimistically to the winter of 1946-47. Nonetheless, coal supplies were of constant concern to civil servants at the Ministry of Fuel and Power and Douglas Jay (Attlee's private secretary until he became an MP after the Battersea by-election in June 1946) acted as a conduit for these fears to the Prime Minister.[38] Jay's concern and Attlee's natural caution led to the recalling of the Ministerial Coal Committee (under Hugh Dalton) on 1 August but the main obstacle to an effective response was Shinwell.

Shinwell did not dispute the potential for a fuel crisis but believed the problems were surmountable.[39] His optimism was based on the assumption that public ownership would create a new spirit in the pits which, with increased recruitment, greater fuel efficiency and more oil burning would get the country through the winter without economic disruption. This attitude climaxed in Shinwell's hubristic statement of 24 October when he declared the only person who did not know of the developing fuel crisis was the Minister.[40] In October and November 1946 there were unexpected power cuts which leads Robertson, in his definitive study of the crisis, to argue 'that at the very beginning of January 1947 (if not indeed by Christmas 1946) the country was no longer merely moving towards a crisis in its fuel and power resources, but had gone over the brink.'[41] At the beginning of the coal winter (1 November to 30 March) rising coal output was being outstripped by demand and coal stocks were 20 per cent lower than in November 1945, so the margin of error was wafer thin and Shinwell's response (forced on him by Attlee) was inadequate. The fuel crisis was a reality weeks before the winter. On 3 January Shinwell finally admitted to a stunned Cabinet the seriousness of the situation.[42] Horner's memoirs contain a graphic portrayal of the crisis:

> They sent a car for me, and we slithered through the icy streets to Millbank. Shinwell had Harold Wilson and Hugh Gaitskell, his Parliamentary Secretary, and some of his officials with him. They told me the situation was critical, and said that unless London had coal by the following Wednesday (this was a Sunday night) London might have to be evacuated. They explained to me that the sewerage could not be worked without electricity, that water supplies were in jeopardy and that, in fact, the life of the capital would be brought to a standstill.[43]

Sir Stafford Cripps was put in charge (7 January) of meeting the crisis, promulgating (13 January) the Cripps Plan (priority of coal for power stations and a 50 per cent cut in power supplies to industry) which was announced but was immediately destabilised by the weather, this in turn was supplanted by the Cabinet Coal Committee (12 February) chaired by Attlee.[44] The seriousness of the crisis for both the coal industry and the government was not lost on the NUM as maximum output was essential to 'defeat the activities of those reactionary forces who are seeking to sabotage the efforts of the Government.' The crisis the NUM argued, was not the result of public ownership but a combination of years of neglect and the weather, indeed public ownership made the crisis easier to resolve.[45] Shinwell was sidelined during the worst of the crisis which by 18 February was stabilising, and by 27 February the worst was over.

Shinwell used the fuel crisis to buttress, ultimately without success, his position in the Labour government.[46] Shinwell focused the NUM's attention on the class hostility to public ownership: 'Does anybody really suppose that this capitalist gang are going to allow us to nationalise where we please and change the order of society without a struggle?' Shinwell compared the fuel crisis ('a very trifling affair') with 'the crisis that is bound to be forced upon us by the determination of those who are no longer in power to wrest power from us at the earliest possible moment, and by any means...our opponents will stop at nothing to clear us out.'[47] At a meeting with the NEC at 10 Downing Street in March, Attlee argued an annual output of 200m tons 'constituted the absolute minimum' for the domestic and industrial market as well as to fuel the export drive. Priority had been given to recruitment and boosting the supply of machinery to the pits, extra rations were allocated to mineworkers, more consumer goods would be directed to the coalfields and, despite shortages of materials and labour, the government would press ahead with the construction of baths and other pit-head facilities. In return, Attlee wanted coal stocks to be boosted to 15m tons at the beginning of the coal year and the NEC agreed to appointment a Special Committee to consider how to achieve this.[48] Horner argued 190m tons of deep mined coal would require 100,000 extra recruits but it was likely only 73,000 would enter the pits and direction would not work (the wartime Bevin Boy experiment had shown this) but if the coal was not produced heavy unemployment would result. The gap could not be bridged with US coal imports because of the dollar shortage so increased recruitment was the only solution but men would only enter the industry 'if the conditions of employment are sufficiently inviting.'[49] The NEC's review of 1946 concluded 4m tons of coal per week had to be raised and 'only when this output is achieved can we really begin to regard the position as being satisfactory.'[50]

On 30 July Attlee called the NUM and NCB to a meeting to discuss production. Significantly, the NCB's delegation made no contribution as the main purpose of the meeting was to persuade the NUM to accept longer working hours. Herbert Morrison warned that 'without an immediate increase in production generally, the country must face all-round cuts in the standard of life of our people. Everything depended on the question of production and the government were satisfied that it all began with coal.' The NUM's members, he pointed out, could not be isolated from any collapse in living standards caused by a coal shortage but the government would trust the mineworkers and not interfere with the Five Day Week or call for the resumption of Saturday working to increase output, instead it sought an extra half-hour on the working day. Increased shift hours were deeply unattractive to the NUM as it symbolised the worst practices of the private owners who had used their victory in 1926 to increase working hours. More injuries were sustained in the last half hour than in the rest of the shift. Attlee presented the proposed increase as a temporary emergency measure lasting at most 18 months to two years and Ernest Bevin agreed the government would consider alternative proposals on condition 'they would not add too much to the cost of coal... The Government did not want to have to face inflation through the price of coal and then have to face the miners with a reduction at a later period.'[51] The NUM and NCB were agreed the basis of policy was the full implementation of the Five Day Week (increased stints, the end of restrictive practices and the dismissal of persistent absentees) but the NUM strongly preferred an emergency additional shift for the duration of the production crisis.[52] Ministers remained convinced that the answer lay in a longer daily shift increased from seven and a half hours plus one hour winding time to eight hours plus one hour winding time but the NUM remained adamantly opposed to increased shift times. NCB opinion was divided but there was an appreciation that the NUM's objection was insuperable and that ministers ought to allow the Board more flexibility in finding a solution.

The Special Conference called to secure approval for extraordinary measures to respond to the fuel crisis heard Morrison make an impassioned appeal. 'We have', he pleaded, 'got to have a new mentality, a new soul in industry. We are in the midst of a transition, if not a peaceful revolution, and if men are going to vote Socialist and want Socialist things, they must act socialistically in their daily life.'[53] Horner conceded there was serious resistance in the coalfields but this could be broken down by patiently explaining to the men the need to meet their obligations. Those who refused were helping the enemies of the nation, of the government and of the miners by giving succour to those who were determined that public ownership would fail. Horner reflected on the tremendous symbolism to the mineworkers of an increase in hours and the NEC were convinced an eight hour shift would not deliver the coal but it did support the proposal for systematic albeit voluntary, Saturday working.[54] Horner's reply to the debate testified to his and the NEC's frustration about the difficulty of increasing production:

> We are the fathers of the five-day week, and we do not want to kill our own baby. We want to keep it. But we must relate it to the situation. We have heard that neither one nor the other is any good – the eight-hours or the Saturday. Well, what is any good? If nothing is any good, then we are sunk; the country is down the drain. If we say, "What is the good of anything? Nothing." Well, just wait until you are starving to death.

If the coal was not forthcoming 'our manpower problems will be solved through starvation and destitution' and the Labour government would be supplanted by a 'reactionary Tory government.'[55] The resolutions were passed unanimously.

The Cabinet conceded eventually that a general increase in shift time was politically impossible and permitted the NCB to negotiate a Saturday Working Agreement. This recognised the NUM's preference for voluntary Saturday working as a temporary expedient and it was agreed that Saturday working would be paid at overtime rates and would not be regarded as part of the normal working week. Ministers had tried to persuade the NCB that the bonus shift should only be paid after six shifts had been worked but as this was tantamount to compulsory overtime and a clear breach of the

FDWA it was dropped. Still hankering after a longer shift ministers secured an Order-in-Council which amended the Coal Mines Regulation Act (1908) to permit an 8 hour shift plus one hour mining time as a local option. Only Northumberland and some Durham pits opted for this. An agreement was hammered out with the NCB on 20 October 1947 but neither voluntary Saturday working or the local option was (in theory) to be introduced until the FDWA was fully implemented and restrictive practices had been abolished. Moreover, the agreement was temporary, lasting until 30 April 1948.[56] To deflect criticism the NEC stressed the Agreement had been negotiated freely, no threats had been made by the government and the Agreement was for the duration of the fuel emergency. The NEC stressed its refusal to concede a longer shift despite considerable pressure to do so but warned the FDWA had to be fully implemented to make up the 14-18m tons lost and the extra output would demonstrate the value and success of public ownership to the community. The agreement was approved.[57] In its Annual Report the NCB noted that 'In many instances, the mineworkers did accept increased tasks. The experience varied from coalfield to coalfield and from pit to pit. Taking the country as a whole, the response to the appeal by the Board and Union for greater tasks was disappointing.'[58] Nonetheless, the increased output was a welcome addition to fuel supplies even if it did not fully resolve the production crisis.

As the summer stock building period drew to a close the NEC recognised the industry was again facing a production crisis. The government needed an increase in production from 4.25m tons to 4.4m tons per week and 13-15m ton increase in 1949 over 1948, with a target of 250m tons. Gaitskell reminded the NUM and NCB that 'nobody had found an effective method, by Government action, of cutting down electricity consumption' so coal output had to increase. Failure to produce enough coal would damage the economy, the NCB had made a £23m loss and there was constant pressure from the NUM for improved wages and conditions but these could not be financed by an increase in coal prices, again the solution was increased output.[59] Bowman argued the NEC 'had a duty' to formally acknowledge that the coal industry would fail to meet its production target by 3m tons. The declaration was circulated widely in the industry.[60] By February 1948 Saturday working had been suspended at 13 collieries in the North East Division and 8 in the North West costing 180,000 tons of coal per week, and 52 collieries had never operated the Agreement. Implementation was a matter for local agreement but all collieries were expected to make up the production lost under the Five Day

Week but in some Areas the custom of leaving the pit when the contracted stint had been completed was continuing.[61]

By early 1949 production problems had forced the NUM onto the defensive. The NCB (supported by the Government) was resisting NUM claims for further concessions and required 'a substantial improvement...in attendance and effort resulting in increased output and reduced costs.'[62] The NUM accepted the seriousness of the crisis but it was under pressure from the Areas not to renew the Extended Hours Agreement (EHA). Gaitskell was uncompromising as the Government's plans called for a 6.5 per cent increase in output but the increase was only 2.5 per cent, a rate which would seriously damage the export drive. To discontinue the EHA 'would be calamitous [but] a failure to meet the requirements of the country must have severe repercussions...and could only result in a serious setback for the nationalised industry and for those whose living depended on it.' After Gaitskell left the meeting the NEC agreed to recommend the extension of the EHA for twelve months and make every effort to solve the problem of absenteeism.[63] Gaitskell's complaint about 'the failure of deep-mined production to meet all the requirements of the country' led to an urgent meeting with the NEC and the NCB in November. Demand for coal was 23m tons higher than in 1945 and 35m more than in 1938 because of full employment and the export drive but in mining manpower and output were falling. Hyndley told the NUM 'it was a heavy blow to them that the coal industry was not able to give the country the coal it needed' but the NUM argued the crisis was foreseeable and more (especially in recruitment) could be done. The NCB was deeply concerned about the drain of trained manpower to other industries and the 300 pits not working the Saturday shift, noting that a 1 per cent fall in absenteeism would produce 2m tons extra coal a year. The NEC argued full implementation of the EHA and improved conditions to attract more labour would substantially increase output and these improvements were a national (i.e. government) task. The NCB was disappointed that the industry could not 'meet a rapid increase in consumption...so that exports had to be cut and emergency measures taken to carry the country through the winter of 1950-51.'[64]

The 1949-50 fuel crisis led at Attlee's request to a meeting with members of the Cabinet with the NUM Executive. Coal stocks had to be kept above 9m tons but efforts to build a buffer stock had failed and American coal had been imported at the cost of scarce dollars. There now had to be 'a great effort' by the mineworkers to keep the power stations running in order to sustain economic growth and the rearmament programme. 'The miners', Attlee warned, 'had it in their power to save the country from an immense loss of production and a great deal of hardship and privation, if an all-out effort were made.' To avoid a crisis output had to be 3m tons above 1949 for the next few months and another crisis like that of 1947 would be disastrous:

> It had needed a long fight to get nationalisation adopted. To have another coal crisis now would give every critic of the chance to say that it had failed. There would be a demand for an Inquiry and for a reversion to the old ways of the past...It lay in the hands of the miners to show that nationalisation could be made a success.

Jim Griffiths, the Colonial Secretary and an ex-miner, reiterated that 'another fuel crisis...would bring discredit upon the Government and the policy of public ownership' provoking massive and damaging criticism from the Conservative Party and the press. Ernest Bevin was the heaviest gun deployed. He warned that the US coal industry was ready to export a vast amount of coal to Western Europe and the UK had been forced to renege on export contracts which would increase prejudice against British coal (and other goods) in Western Europe. Export earnings from coal paid for food and other supplies and a fuel crisis, coupled with worsening austerity, could jeopardise all achieved since 1945 and lose Labour the next election:

> There was a great new electorate growing up in the dormitory areas around London. They were a powerful factor and renewed shortage of coal might turn them against the policy of nationalization. It was up to the industrial areas to do their utmost first to maintain our vital exports of coal, secondly to preserve the hard won balance of our economy which we had achieved and thirdly to keep the confidence of the other sections of the community.

Under this battering the NEC expressed their willingness to fully co-operate 'but there were certain points on which the success of that co-operation depended' and the main problem was falling manpower so wages had to be increased. Philip Noel-Baker (now the Minister of Fuel and Power) warned that these were matters for the NCB and the government 'would not intervene' but would consider the NUM's claims 'sympathetically'. The immediate task was increased output. Bevin backed up Noel-Baker arguing 'it would not be right to allow the present crisis to become a matter of bargaining; if the miners would make the effort needed to overcome the crisis the government would not take advantage of that fact and the miners would have created a public opinion in their favour which would be worth a great deal for them.' The miners, Bevin warned, should not expect too much, too quickly and concluded '[T]here could not be large rises in wages as well as improvements in the social services unless there was at the same time a great increase in output.' No commitment was made by the NUM but the Cabinet's message had been received.[65]

Despite stressing the government's 'sympathetic' attitude to the NUM's claims the NCB were reluctant to make concessions on pay unless the union did likewise on production.[66] Production difficulties were nearly as bad as in 1947 but the political situation was even more dire 'for the day there ceases to be the absolute minimum of coal necessary, that is the day this Government will fall, and the Government who will come instead will be a Tory Government.'[67] The NUM did deliver (just) the extra coal sought by the government but it warned ministers of growing resistance in the coalfields despite the very real desire of the union and workforce to sustain the government.[68] Horner's impassioned justification for increased output was rejected by a South Wales delegate who insisted 'it is a myth...to believe that the mining industry is going to save the government in the next Election.'[69] This pessimism was reflected by the Joint Production Committee where the NUM's representatives confessed they 'did not see how, as matters were going, a better result than two to two and a half million tons could be achieved.' By the end of March 1951 the 3m ton target was in sight although throughout 1951 coal stocks remained precarious.[70] In November 1951 Houldsworth told CINC that stocks of 19m tons would mean no anxiety, 18m tons represented an even chance of getting through the winter but in November stocks were 16.97m tons and 1m tons had been lost in unofficial industrial disputes. Robert Hall of the Cabinet Economic Secretariat confessed that 'The worst thing is coal. The labour force drops steadily...because of the boom in investment industries. We could export all the coal we could dig to Europe...But for all our planning we cannot get 0.1 of 1 per cent of the labour force to move.'[71]

Only after Labour was defeated in the 1951 General Election did the coal supply situation appear to ease. By mid-1952 the post-war production crisis appeared to be at an end, so much so that one delegate argued 'the [NCB] may well toughen up now that they think they have a better manpower position and increasing output.' By the Autumn coal stocks stood at 17.75m tons and 'It looked therefore as if the industry could face the winter without the atmosphere of crisis which had prevailed so frequently in recent years...'[72]

The Politics of Collective Bargaining

Central to the NUM's policy for more output was increased pay, but the strategic position of 700,000 mineworkers and coal in the political economy made their pay a highly sensitive issue. The inflationary consequences of the 1944 full employment commitment deeply concerned ministers and civil servants.[73] The largest single cost in the industry was pay, so wage increases would affect coal prices and have a 'knock-on' effect in other industries with dangers of inflation, employment and economic recovery. By 1949 the annual pay round was an established feature of industrial relations and, expecting inflation, bargainers built it into their wage claims. This created the conundrum of how to stabilise wages whilst preserving free collective bargaining, the touchstone of British trade unionism. Government was keenly concerned with pay bargaining in the coal industry and the public sector wage bill was an attractive, if politically sensitive, regulator.

Ministers made it abundantly clear to the NUM that the government would not concede substantial increases in coal prices in order to pay increased wages unless the NUM secured full five day working and ended unofficial stoppages. The very first wage bargaining round under public ownership revealed the tensions and problems that were to characterise post-war wage bargaining. Faced by an unofficial strike at 68 Scottish pits over wages a furious Hyndley warned the NEC that he had 'felt strongly disposed' to end negotiations but refrained as the NUM's negotiators were already on their way to London. He also told the NUM that despite its conviction that 'the nation must pay the necessary price for the coal it required' he would not seek an increase in coal prices. The NCB rejected a general price increase as the Board was not covering its costs and coal prices could not be raised without damaging the rest of the economy whilst a subsidy 'would mean nationalisation had failed.' The NUM's object was to secure 'the least possible [wage increase] which would restore satisfaction to the ranks of the men and attract new men to the industry.' A compromise was eventually reached 'but only after very considerable hesitation' on the NCB's part which was relying on the NUM 'to do everything possible to make the men realise that they must now put their backs into their work and secure the increased production which was so vitally necessary at the present time.'[74] The NEC concluded further negotiations were unlikely to produce significant concessions and its acceptance of the offer was criticised (for example, by Abe Moffat) as contrary to Conference's instructions but, as pointed out, 'substantial' had not been defined. Attempts to re-open negotiations were overruled by Lawther.[75]

The negotiation of this first wage claim was significant as the NEC appropriated a very wide degree of latitude in interpreting (admittedly vague) Conference resolutions in the interests of defending both the Labour government and public ownership. Another feature of the 1947 claim was the importance attached to the 'national interest' as a factor in pay bargaining. The NCB frequently cited CINA's requirement that the industry produce the coal the nation needed at an economic price and, as the foremost advocate of public ownership, it argued that the NUM was obliged to help deliver this objective. However, the state of the NCB's finances could not be ignored. Although the NCB was not required to make a profit it was expected to cover its costs on an average of good and bad years, and the NCB argued that the mineworkers, as partners with management, must take the NCB's finances into account when formulating a wage claim. Despite the claims of the Board and government, there *was* a direct political input into the collective bargaining process. Whilst the Cabinet did not intervene overtly, its determination to keep wage inflation and fuel costs down was well known to both the NUM and Board and was an unavoidable factor in wage negotiations which as a result of public ownership were now more political and more complex than under private ownership.[76] However, a body of opinion in the NUM questioned the extent to which political considerations should determine union policy.

On 4 February 1948 the government published the White Paper *Statement on Personal Incomes, Costs and Prices* which marked the beginning of Labour's formal wage policy. The NUM considered the TUC General Council's statement on 11 March and agreed to support it on condition that free collective bargaining was preserved, that increased output would be rewarded, low pay addressed, wages in undermanned and essential industries could be improved to attract labour, and skill differentials would be safeguarded. The NEC meeting on 1 April agreed to support the government's objective of wage stabilisation (approved by the TUC General Council on 18 February) on condition that it was accompanied by price control and that it would support the TUC's policy at the Conference of Trade Union Executives on 24 March, two weeks before the Budget. This conference approved the policy by 5.4m to 2.0m. This decision provoked vocal opposition in the NUM from Kent and Scotland who protested that the NEC's decision was taken without consulting the membership.[77] The NEC's actions were influenced by a desire to support 'their' government rather than defend their members' interests and disquiet over wages emerged forcefully at the 1948 Conference.

The wages debate (on resolutions from Scotland calling for a new wages structure and from Nottingham calling for workers to be paid the rate for the job) was complicated by the NEC's decision to support the TUC's policy. Abe Moffat stressed the discontent of the low paid which could only be resolved by a new wages structure and not to do so endangered public ownership 'because you are carrying forward into a nationalised industry a wages system that was rotten...creating conflicts in every pit and in every district in the British coalfield.'[78] Scotland also moved the reference back of the NEC Report dealing with the NUM's support for the TUC on the grounds that in so doing the NEC had openly defied a Conference decision and the matter should have been referred to a Special Conference. Naturally Lawther refuted this, arguing the NEC had been asked by the TUC to consider the policy as a matter of urgency and to do so in line with the needs of the mineworkers as part of the wider labour movement. Sam Watson argued the NEC acted according to NUM Rules as it 'took a decision in relation to a very important issue confronting this country, and in so doing they accepted Executive responsibility.' The reference back was defeated by 494,000 to 161,000.[79]

The NEC were well aware of the sensitivity of this decision and the political risk it involved. Horner insisted that the NUM had achieved much in a brief time, partly through its own strength, partly through the justice of its case and partly with the help of union allies and the government. The NUM's strategic political and economic location could not be ignored:

> You must appreciate that a nationalised industry, situated in the centre of and surrounded by private enterprise, is largely conditioned by that which obtains in private enterprise. You cannot insulate it and make an oasis of a nationalised industry with no association whatever with surrounding industry. Every time we make a move, every time we make an effort to change conditions in the mining industry, we are brought face to face with the complications and repercussions which will arise from that change in the surrounding privately-owned industry.

Horner urged the delegates to remember their history, to remember what had happened after 1918 and compare this with what happened after 1945; the NEC 'were determined that this time we should emerge from this disastrous war in a position in which we could ensure that there would not be the widespread poverty and distress which followed the earlier war.'[80]

In his ministerial address Gaitskell had warned delegates that improvements in pay depended on increased output and he reminded the NUM that the NCB had made a £23m loss in its first year. This linking of pay to the NCB's finances greatly worried the NEC.[81] Growing grassroots discontent over wages and over aspects of public ownership, forced the NUM leadership to reconsider their strategy. Supporting the government while the NEC ignored grassroots hostility risked severe internal conflict and although the NUM was committed to making public ownership work, increasing numbers of NUM members feared this was being done at their expense. Support for Labour's pay policy was originally presented as being in the membership's real long-term interests by preserving the Labour government, maintaining full employment, and combating inflation in order to preserve the real value of the wage packet. As restraint was increasingly perceived by many mineworkers as not being in their immediate interests, the NUM leadership strove to integrate the union and its members by other means. One method (discussed in the next section) was to attack the Communists in the NUM; a second strategy, which evoked powerful memories, was to warn that if the Labour government fell the result would be a Conservative government. A third strategy was to manage (or manipulate) the Conference agenda via the Conference Arrangements Committee to block or neutralise unpalatable resolutions, and the NEC majority could prioritise Conference resolutions in order to maximise the negotiators' room for manoeuvre.

From late 1948 the NCB was concerned about costs and, in response to clear signals from government, its resistance to concessions not directly related to increasing output stiffened. The NRT rejected the NUM's claim for extra holidays (for example) on highly significant grounds, 'the present position of the country as a whole, the necessity for the industry to pay its way, and the economic effect of a rise in the price of coal, [mean] that the claim cannot be safely granted at the moment.'[82] This resistance was blamed on the government by many in the NUM and appreciating the danger of this the NEC moved quickly to scotch this asserting 'it had always been considered inadvisable that the Minister should have any authority to intervene on a question of wages or conditions of wages.' The NEC stressed that Gaitskell accepted non-involvement.[83]

At the 1949 Conference four resolutions were composited calling on the NEC 'to endeavour to secure a substantial increase in minimum rates for all lower-paid workers.' This was a politically sensitive resolution given the depth of the country's economic crisis which was to culminate in the September devaluation. Willie Pearson argued remedying the plight of the low-paid was a test of the NUM's socialism. He agreed the country was in a deep crisis and so 'We either meet it in the old capitalist way, which is at the expense of the working class, at the expense of their wages...or the socialist way'. The 'deep discontent' in the pits was a warning and Pearson cautioned the NCB 'that if they are tough with the miners, never forget that the miners can also be tough, and very tough' and he concluded that if the NUM did not fight for the lower-paid they would do it for themselves 'then along will come that mysterious agitator whom we always find when workers are struggling for their rights'.[84] Jim Hammond made a broader point. The NUM's members were affected by prices rises which were the result of the market, mineworkers should therefore obey the dictates of the market and 'If we have to live in [sic] capitalistic economy, if things have got to be dependent upon competition either in the supply of labour or commodities...all right, let competition obtain as far as labour in the pits is concerned, and let wages be sufficient to bring the labour that this industry needs.' Low pay in the pits was because 'capitalism wants to enjoy the benefits of cheap coal from this industry, and they see nationalisation as becoming an efficient social service for the rest of capitalism.'[85] Lawther's response was to seek the composite's remission to the NEC. The only way low pay could be resolved he argued was by an increase in the price of coal and this would not be sanctioned by government. Lawther accepted there was a universal agreement in the NUM that low pay had to be ended 'but I say to you, in this hour of the nation's history, and the position which the Chancellor of the Exchequer has explained to Parliament, that it is not playing the game with your own Government, and you are forcing a situation that will lead inevitably to a clash.' Horner presented the NUM as caught between the long-term necessity of restructuring the industry's wages and the need to respond to the immediate crisis. Low pay could not be remedied unless and until the economy stabilised and, meanwhile, the NUM intended to use cost of living agreements to direct money to the low paid.[86] The Composite was remitted unanimously.

In September 1949 Sterling was devalued by 40 per cent. The TUC prepared plans for an absolute wage freeze but its draconian proposals were rejected by affiliated executives. In November the NEC received a letter from the TUC General Council concerning a meeting of the Special Economic Committee with ministers to consider further measures to control inflation during the 'economic emergency'. The TUC urged all unions to postpone current wage negotiations until the General Council was able to deliver a further statement on pay guidelines. The NUM's representatives on the General Council urged the NUM to agree (thereby postponing negotiations with the NCB) until the Conference of Executives was held in the New Year. The NEC agreed despite growing opposition from the coalfields.[87] The General Council suggested union members forego any wage increase as long as prices did not rise above 5 per cent and low pay was to be remedied only in exceptional cases. The TUC's case was that inflation 'must be countered by vigorous restraint upon all increases of wages, salaries and dividends', restraint should be voluntary and 'whilst it is the responsibility of the Unions themselves to operate the wages policy, Unions nevertheless must pay regard to the realities of the economic situation in framing their policy and act loyally in conformity with the policy.' The NEC disliked much of the policy but believed the alternative was deflation and unemployment and after considerable and forthright discussion it agreed to recommend acceptance of the TUC's policy.[88] The NEC hoped the Labour government's plight, coupled with the policy's voluntary and self-policing nature would persuade delegates to accept it. However, delegate after delegate came to the rostrum to argue that whatever the policy's merits they were unlikely to be able to convince the men in the pits. Lawther bluntly warned delegates 'whatever may be your decision, we shall have to face a General Election. Either we stand up to this now, or we allow a breach within our ranks. Make no mistake, the choice is between what the Labour Government and the Trade Union movement is able to do, or the opening of the floodgates of reaction.' Responding to a question on how the acceptance of the TUC's policy would affect the low paid Lawther was unequivocal: 'I cannot see how we could accept a policy in relation to wages restraint owing to the general economic position and in the next breath go along and say, "We want something here that will cost millions"...'.[89]

Horner moved the resolution supporting the policy on behalf of the NEC with a notable lack of enthusiasm, signalling his personal dislike of the policy in a short speech stressing his duty as General Secretary to represent the majority NEC view. Delegates who spoke against justified their opposition using two familiar arguments: this was a crisis of capitalism and it was not the workers' responsibility to resolve it, and the NEC were betraying the low paid.[90] Bill Paynter (South Wales) claimed 'we are actually on the threshold of a transformation in our policy.' The NUM had been making progress in recent years but the NEC was now proposing to halt this forward march by urging its members to help resolve the crisis of capitalism. Paynter drew a parallel that passing the resolution would not help Labour, the unions or the working class as the capitalist system would remove a Labour government unless it followed capitalism's remedies. If the resolution passed there was still no security or guarantees for the miners or anyone else, especially 'if we are going to pursue a policy that is identical with the policy of the Tories in this situation.'[91] A Scottish delegate, a faceworker on permanent night shift, warned 'the Scottish miners will never accept the policy as propounded by the platform' arguing the NEC's advocacy of wage restraint demonstrated their lack of contact with their members. Jim Conway (Yorkshire) agreed external events had affected government policy 'but it is even more remarkable how the external events of the last four years have entirely failed to play havoc with the profits of the industrialists' and concluded that even if the Special Conference approved the TUC's policy the men in the pits would ignore it.[92] Delegates from moderate coalfields warned 'we shall have considerable difficulty in the coalfields...It will not be easy convincing Notts miners that we must put the interests of the Government or the TUC or the [NEC] before their interests.'[93]

Advocates of wage restraint stressed the need to support the Labour government because they were trying to avoid the deflation and unemployment which had cost the miners dear in the 1920s and 1930s; second, this support would also help prevent the return of the Conservatives; and third, 'Because so much has been done for us'.[94] Jack Besford (Northumberland) wondered what would the NUM do if it ignored the TUC and pushed the wage claim, which was then rejected by the NCB? Should they refer it to the NRT? If the NRT rejected the claim should the NUM strike? The miners and the working class had gained much since 1945 'and it is up to us to do something for the Labour Government' because the imminence of a General Election and the possibility of a Conservative government 'transcends all questions of immediate increases in wages.'[95] The NEC's resolution to support the TUC's policy was approved by 406,000 to 173,000 by the Special Conference, but the Areas rejected the Special Conference's recommendation by a majority of 371,000. The NUM therefore opposed the TUC's policy at the Conference of Trade Union Executives which rejected further pay restraint so killing the government's incomes policy. The NUM's rejection of the wage policy demonstrated that no union could restrain or control membership discontent permanently so the NCB and government had to be satisfied with the amount of cooperation they could secure or which could be delivered by the NUM's leadership.

Demands for immediate action by the NEC on behalf of the low paid continued. Abe Moffat argued that as a matter of principle the NUM should fight for the lowest paid and if this brought it into conflict with the

NCB and government, so be it.[96] The NEC recognised the strength of feeling and announced its support for the composite's *objective* but questioned how it should be realised. The NEC had begun work on revising both the wages structure and the position of the low paid but aside from the complexity of the problem, there was a political dimension which the NUM could not ignore. The Coal Board was a 'Government agency' which meant 'it cannot easily pursue policies which are contrary to the policies of the Government especially if those policies are the accepted policies of the official Labour Movement.' The NEC was trying to find a wages formula which would not give most to the highest paid and recognised that to do otherwise meant 'the coalfields would explode. It would be impossible to hold the situation.'[97] Both the composite and the NEC's statement were accepted by Conference.

By 1950 the NCB was arguing consistently in negotiations that neither it nor the NUM could ignore the industry's finances or the statutory requirement to manage the industry economically and efficiently. Coal was now three times more expensive than in 1938 and the NUM's members had made substantial gains since 1947 as wages had increased by 35 per cent.[98] Faced by deadlock in negotiations the next stage was to refer the NUM's claim to the NRT. To the intense disappointment of the NUM the NRT's Twenty-Third Award did not concede the NUM's case. Only £3.5m was made available for a wage increase raising 'the prospect of disturbance in the most highly producing districts' but the NUM agreed do all it could to implement the Award (as it was bound to do under the Conciliation Scheme) 'but they could not give guarantees of peace or any assurance of cooperation in the coalfields.' The NCB rejected an NUM suggestion that it ignore the NRT Award as this would destabilise the conciliation machinery and argued the NUM should 'be anxious to co-operate with Board in preserving the sanctity of the Award and conveying to the men in the coalfields that an Award...should be accepted with goodwill.'[99]

The Twenty-Third Award was a significant event in the post-war politics of the coal industry. The reality of collective bargaining in the coal industry was driven home at the Special Conference called to discuss the NCB's wage offer. Horner's report on negotiations stressed that the NCB, buttressed by government policy, was immovable and he cited the emphasis placed by the Board at the October 1948 meeting where Gaitskell had made future concessions to the NUM conditional on increased output. The NCB were not independent bargainers but Horner believed they were negotiating in good faith with the NUM. Consequently, 'we have emerged with about half the sum of money we set out to get in the beginning [and] we are recommending acceptance...because we have no alternative.'[100] Although Lawther had ruled out of order a wider debate on the conciliation machinery, several delegates nonetheless raised the question of the NUM's strategy. Abe Moffat, for example, suggested events had shown that an Area vote should decide whether an NUM wage claim 'is to go to arbitration or whether the miners have to use their organised strength.'[101] This raised the wider question of the NUM's response if the Areas voted again to reject an NEC or Special Conference recommendation, and the radicalism of Moffat's call to consider the use of industrial action should not be underestimated. This was taken up by Paynter who asked, 'Are we going to accept the philosophy that half a loaf is better than no bread, or are we going to say we are going forward to realise the substantial increase we have been instructed to obtain?' He further argued that arbitration was part of the government's policy of wage restraint which had been rejected by the coalfields in January and 'the logic of our experience today, as well as the logic of our experience in the past, is that...we have to rely on our strength as an organisation; we have to rely on our bargaining power'.[102] Knowledge of the NUM's disruptive potential was however itself a powerful restraint because 'Immediately we used our strength, because we are key-workers in the national economy, it would be immediately followed by the steelworkers, the railwaymen and the transport workers, and out of this jungle warfare we would have chaos.' The offer had been negotiated using machinery which the NUM had helped create in a publicly owned industry the mineworkers had long sought and one of the main justifications for nationalisation was constructive industrial relations. The NUM was obliged to make the system work.[103] Government resistance to further wage increases in the coal industry remained strong until it left office. In January 1951 the NUM was again warned against attempting to exploit the present crisis and if they did not the government would be sympathetic. However, as in the past the NUM was warned that their wages as well as full employment and the welfare spending depended on increased output.[104]

The Coal[d] War

Traditionally the internal politics of the NUM were free of a sharp Communist/anti-Communist cleavage. After June 1941 the CPGB were staunch advocates of workforce discipline and maximum production, and the 1945 NUM Conference approved by 479,000 to 166,000 a resolution supporting the electoral unity of the Labour Party, the Communist Party and all progressive organisations. By 1946 however, the NUM opposed closer Communist Party links with Labour and the NUM's internal politics soon became characterised by a sharp division between Labour and Communists.[105] The breaking point was Britain's decision to support the Truman Doctrine (March 1947) whereby the United States undertook to contain Soviet expansion and the institution of the European Recovery Programme (ERP, or Marshall Aid) in June 1948. In April 1947 NATO had been inaugurated and in the autumn the Communist Information Bureau (Cominform), widely seen as an instrument of Soviet subversion, was instituted. Although not invited to participate in its creation the British Communist Party aligned itself with the Cominform and by December 1947 the CP was accused of fomenting industrial unrest to undermine the production drive. The Cold War intensified with the Communist coup in Czechoslovakia (February 1948), the Berlin Airlift (March 1948-June 1949) and the outbreak of the Korean War (June 1950) which led to a huge British rearmament programme. Coal's importance in Britain's recovery and rearmament and Britain's role as the European core of the nascent Western Alliance ensured that the NUM which contained many Communists and their sympathisers, became an important Cold War battleground. Rigid Cold War divisions reduced the NUM's ability to respond flexibly to a changing post-war political and economic environment.[106]

Initially the NUM was reluctant to be drawn into the wider ideological battle. Lawther, for example, criticised the growing gulf between the wartime allies and whilst welcoming the Marshall Plan expressed concern about Britain's growing reliance on the United States:

> We are not unmindful of how quickly the richest nation in the world ended Lease-Lend when the Labour Government was elected, as a threat to it right from the start. We notice what American monopolists are doing to try and crush their own Trade Union Movement; we do not close our eyes to the racial hatred and discrimination in what is alleged to be God's Own Country.

Lawther condemned the growth of anti-Sovietism in Britain and the USA and called on democratic forces everywhere to resist the drift to an atomic war with the USSR, arguing 'no British miner or other conscientious worker would produce any materials for such a war, and the sooner the warmongers understand that, the better.' Britain should declare 'its firm intention to stand alongside the Soviet Union and other peace loving nations where ever they may be, then no other nations in the world dare talk of a new war.'[107] However by January 1948 the atmosphere had changed.

Arthur Horner, the NUM's General Secretary and a Communist, became the focal point of an attack intended to isolate Communists in the NUM.[108] A special meeting of the NEC before the 1947 NUM Conference at Margate had criticised a personal statement made by Horner on the coal industry on behalf of the CP. The NEC reiterated that officials should not make statements unless they were approved by the NEC and were in conformity with NUM policy.[109] In January 1948 the Scottish Area protested against leaked reports in the press that Horner was being heavily criticised by other members of the NEC as designed to undermine his credibility and discredit Communists in the NUM. The Scottish Area called on the NEC to repudiate these reports which led to a counter accusation concerning a letter from Willie Gallacher MP, communist MP for West Fife, which had criticised Lawther and the Labour Party NEC. An acrimonious and politically divisive Conference debate was stopped by moving 'next business.' One month later the NEC sponsored a resolution at the Labour Party Conference expressing 'loyal support' for the government's foreign and domestic policy, welcoming the offer of Marshall Aid.[110] Gaitskell noted in his diary the CP's changed attitude to the production drive but also that the NUM and coal appeared exempt from the changed line. Nonetheless, he thought this change would complicate the NUM's internal politics and Horner's position and 'The view of both the NCB and Lawther is that he will have to make up his mind as to which side of the fence he is going to stand.'[111] By Autumn 1948 the split between the Communist and non-Communists in the NUM had become explicit and 'voting in their Executive is on strictly Party lines.' The non-CP leaders, Jim Bowman, Sam Watson and Ernest Jones, were briefed privately by Gaitskell on government policy and encouraged to resist the CP so 'the politics of the thing from now onwards will be an alliance between myself and them behind the scenes.' This new alliance was tested at the 7 October 1948 meeting, one of whose objectives was to reduce the CP's influence over production by creating a NUM/NCB Joint Production Committee. To achieve this Gaitskell 'picked out one or two lines I knew the Communists were running, e.g. subsidy of the industry, higher prices, etc., and ruled them out completely. All this is bound to sharpen the conflict within the Union and, of course, we run the risk of local trouble with the CP at any time.'[112]

Between 11 and 15 October 1948 Horner was a visitor to the Twenty-Seventh Confédération Général du Travail (CGT, the Communist union organisation) conference in Paris. Here he made a speech which compared the British strikes of 1921 and 1926 to the French miners' strike which was widely interpreted as a Soviet inspired attempt to destabilise the French state and economy. 'No British miner', Horner told the CGT, 'has authority to speak against the French miners' strike, and if he does so it is entirely unofficial', declaring that 'If we had such conditions as you we should also be on strike' and 'that inspite of propaganda that when the British miners know the full facts of the French struggle they will rally to your support.'[113] On 28 October the NEC discussed at great length Horner's statements on the French miners' strike.[114] These statements had been repudiated publicly by Lawther as contrary to NUM policy and the NEC confirmed this repudiation warning that any further recurrence would not be tolerated. It appointed a Sub-Committee to gather further evidence.[115]

On his return Horner made a further statement at Northolt airport. He presented Lawther's criticisms as part of a concerted political attack designed to reduce opposition to a war with the USSR: 'they have prepared for war and they think it will be very inconvenient to imprison me if at that time I am in the position of General Secretary.' 'They' were the government and union right wingers.[116] Horner's protestation that he was speaking in a personal capacity and that he was the victim of a newspaper smear campaign was rejected by the Sub-Committee. His charge that 'right wingers' were preparing for war was dismissed as 'a slander without foundation' and if the government thought it necessary to arrest Horner his conduct would be the 'deciding factor' and the government would be answerable for its actions. Horner's charge was 'ridiculous and can only be attributed to egotism on his part.' The Sub-Committee wondered why Horner gave such prominence to the poor conditions of the French miners but said nothing about the worse conditions of Polish mineworkers and that Soviet-controlled Poland increased its exports of coal to France whilst the French miners were striking:

> The French and British Communist Parties are far more concerned with creating the greatest possible amount of confusion and chaos with a view to sabotaging the efforts of the governments of western Europe towards recovery than they are with the conditions of French mineworkers. International communism would regard it as a crime to suggest that the miners of a country coming within the sphere of Communist control should take action in support of the French miners and thereby interfer with the economy of their country. To them, however, it would not have been a crime had the British miners taken such action, action

which would have seriously impeded the efforts of our Labour Government to rehabilitate Britain's economy.[117]

The Sub-Committee concluded that the French strike was, on the orders of Cominform, designed to subvert Marshall Aid, and that the British and French CPs took their line from Cominform. The Cominform's second object was the destruction of democratic socialist governments and Horner was promoting Cominform policies because his statements were 'in line with the policy of the British Communist Party which is to weaken the confidence of the workers in the Labour government.'[118]

The Sub-Committee described at length the benefits received by British mineworkers from the Labour government and noted that Horner had been party to all these discussions especially those designed to increase the production of coal. The Sub-Committee noted that it was 'being suggested in certain quarters' that the Labour government and NCB were preparing to attack working class living standards via pay restraint but the Communist Party sought to lower the workers' living standards by 'seek[ing] to place a brake on production.' Horner was condemned as dishonest and hypocritical because he had exhorted mineworkers to produce the coal the country and Labour government needed. 'It is, of course', the Sub-Committee continued, 'difficult to understand how Mr Horner can reconcile these exhortations with the present policy of the Communist Party, the Executive Committee of which he is a member, or with the Editorial policy of the Daily Worker for which he, as a member of the Editorial Board, must accept some responsibility.'[119] The report noted with heavy irony that the Communist Parties of Eastern Europe were constantly exhorting workers to ever greater production efforts so why was it incorrect for a Labour Government to do the same thing?

This was a devastating attack on Horner whose defence was flawed by the contradiction inherent in his position as NUM General Secretary and a leading CP member. In his memoirs Horner portrays this episode as a crucial juncture in post-war labour politics, suggesting the Executive was under pressure from the British and American governments to deal with the Communist threat in the NUM. Horner was under considerable personal strain at the time as his mother was ill, he was suffering from 'flu, and had been under constant press attack because of his CP membership so Lawther's attack on him was the last straw ('I really let myself go'). The Sub-Committee had aspects of a kangaroo court as it met infrequently and submitted its report to the NEC without any discussion. Horner was not called to give evidence in its defence and its report went far beyond the CGT controversy by discussing the role of the British Communist Party even though its remit did not mention the Party. Abe Moffat was incensed by the report and complained forcefully to Sid Ford (the NUM's Administrative Officer) who blamed Sam Watson for the report's strident tone. Moffat then warned the NEC that he objected to his name being attached to a report with which he profoundly disagreed and threatened legal action. The NEC recognised that events had run out of control and noted that while Moffat 'accepted the factual statements contained in the report, he dissociated himself from its political aspects.'[120]

The Sub-Committee, chaired by Sam Watson, reflected the growing bitterness of labour movement Cold War politics. At the time of the Sub-Committee the NEC received and agreed to circulate a letter from the TUC condemning Communist influence in the trade unions and the NEC was to purchase and distribute copies of the TUC pamphlet *Defend Democracy* throughout the NUM.[121] Horner saw these events as indicative of the United States' growing influence in British politics as the Sub-Committee's report reflected faithfully the Labour Government's support for ERP, the USA, and the Cold War. While in France Horner had been told of pressure from the US Secretary of State, General Marshall, on British trade union leaders and believed this was the reason for Lawther's political shift:

> Some time afterwards...I was travelling in a car with Will Lawther...and he mentioned quite casually that he and other members of the TUC General Council had travelled to Paris, without passports and without French money. They had gone in a private plane to meet General Marshall, and there they had been told that they must discourage the French miners' strike or the Americans would stop Marshall Aid.[122]

A Communist as General Secretary of one of the key trade unions in Western Europe worried the Americans.

In July 1947 Hugh Dalton lunched with Sir W. Stephenson who reported 'that the Americans are very frightened of Communism here and of Horner and Cripps in particular.' On the following day Stephenson brought 'Wild Bill' Donovan (the founder of the OSS, the precursor to the CIA) to see Dalton who regaled Donovan with:

> a series of ancient stories about miners; their character, the record of the private mine *owners*, the personalities of the National Coal Board, the total *political* weakness of our Communists, their relative weakness *industrially* – with an excursus on Horner who seems a bit of a bogey-man in the US...I told Donovan

that the one thing we wouldn't stand was any attempt by the US to tell us how to run our politics.[123]

As Minister of Fuel and Power Gaitskell entertained, with Fergusson (his Permanent Secretary) and Lord Hyndley, a number of 'very tough' Congressmen in order to ease their fears about Communism in Britain. Although all went well Gaitskell thought 'It is pretty intolerable to accept patronising comments from people who are quite so odious, [one delegation member] is the chairman of the Congress Appropriations Committee and will have a big influence on the Marshall Plan.'[124] The coal production crisis was seen by many Americans as symptomatic of the failure of socialism. Robert Hall, the Director of the Cabinet's Economic Section, recorded American comments 'that our failure on coal showed that nationalisation was a disturbing factor and that if we nationalised steel, output would fall and US help to Europe would need to increase. Why should they pay for our experiments, especially when they dislike the objective?'[125] US suspicion of the Labour government is well documented so an assault on domestic Communists was a valuable testimony of the Labour government's 'soundness' but there is very little hard evidence of the CP or its members fomenting unrest as opposed to articulating grievances.[126] The press campaign against Horner continued and suggestions from Scottish and Welsh members that the NEC protest were rejected on the grounds that if these were smears Horner was free to use the libel laws. The campaign of vilification against Horner 'helped to frame a "problem": communist trade unionists deceived the men they represented. Hence, strikes could be interpreted not as expressions of felt grievances or class feeling but as the machinations of traitorous leaders.'[127]

CP statements on the coal industry such as Harry Pollitt's *Plan For Coal* (July 1947) criticised the NUM leadership for failing to push for the industry's full socialisation and much improved wages and conditions, charging that public ownership was intended not to help build socialism but to produce cheap coal as a subsidy to private capitalism. In early 1948 Pollitt wrote 'It is in the factories and branch rooms that the workers can best be mobilised and their mass pressure exerted on the government.'[128] Such temerity could not go unpunished. Lawther's 1949 Presidential Address contained a blistering assault on 'the most foul, malicious and venomous attacks' of the Communist Party on the Labour government and unions. The CP's machinations were condemned as responsible for unofficial strikes in the coal industry and 'if you have political views contrary to the Labour Government, and the sooner they are replaced by a ruthless, inhuman Tory Government the better, then I suppose such "quisling" actions of unofficial stoppages can be understood.' Lawther reiterated his belief that in return for public ownership the mineworkers were under a moral and political obligation to produce the coal the country needed and in return for increased production they would receive many more benefits from the NCB and government. The CP's disruption of the production drive threatened these conditions and the government's survival.[129]

As the split in the international trade union movement widened, the South Wales Area wrote to the NEC asking it to do all in its power to heal the breach in the WFTU between pro-US and pro-Soviet movements. The NEC's position was that the movement's policy had been determined by the 1948 TUC which was to put the issue into the hands of the General Council for a decision. The General Council had sought the suspension of the WFTU for twelve months but if this was refused the TUC would disaffiliate so the NEC decided to take no action.[130] The growing political divide could be seen in the debate on the WFTU in a resolution moved by Bill Paynter.[131] A Scottish delegate argued that 'The British miners must make it quite clear that they are against the setting up of puppet splinter organisations based on American capitalism and supported by the American Federation of Labour.'[132] The NEC was determined to see this resolution defeated arguing that the blame for the split lay with the USSR and its allies not the West, 'Was it our Government that sank a battleship in the Corfu Channel, or that brought down that aeroplane on the Gatow aerodrome? Was it our Government that blockaded Berlin?'[133] Bowman argued if the NUM passed this resolution it would be opposing TUC policy and warned that the WFTU was a Cominform auxiliary:

> The Cominform issued its instructions to the Communist parties of the world to attack the British Labour government and the French Labour Government, with particular emphasis on Bevin and Attlee, and from that day onward the new alignment of forces in the World Federation has made it impossible for us to continue to work in that body.[134]

The South Wales resolution was defeated with 166,000 in favour, and 471,000 against. It is worth noting that 20 per cent voted in favour of the resolution.

At the 1950 NUM Conference Lawther added another charge against the CP when he complained that a 'destructive influence is at work inside our own movement. It takes the form of a persistent misrepresentation of its policy and a disparagement of its achievements in home and foreign affairs.' CP parliamentary candidates challenged the NUM's candidates which was 'a right which in every Communist country is denied to those who are not Communists.' Worse than this Lawther argued, was the CP's efforts to undermine the workers' morale and 'All this sort of thing, going on day after day in the factories and the mines, the shipyards and the docks has been having an effect upon the attitude of many trade unionists and wage earners generally towards the Labour Government.' Labour's performance in the 1950 General Election was, in part, the result of Communist tactics.[135]

Lawther's problem was that by the 1950s there was growing resistance to government and NUM pressure for more coal and for pay restraint to keep costs under control. Increasing numbers of mineworkers suspected their interests were being sacrificed to those of the government and NUM leadership. For example, Paynter condemned the 'tendency, based on the achievements inside the country, to regard those achievements with a self-satisfaction approaching a belief that we have now reached the goal of Socialism in Britain.' Socialism had not been achieved because 'the philosophy of Marx is rejected. Socialism based upon the principle of working class power in the economic, social and political life in the country is rejected. And I say that if that philosophy is allowed to predominate..., then we will be following...the policy of continuing Capitalism in Britain.' Challenging Paynter, Sam Watson declared Labour supporters 'do not seek even the qualified support of Communists in this resolution, either you were for the resolution or you were not.' The Labour government had combined 'both bread and freedom...there can be no Socialism without liberty, and there can be no liberty without individual freedom. We reject the theory of totalitarianism and dictatorship, and we rejected the theory of Marx as taught by Stalinism and the satellite countries.'[136] A resolution from Scotland calling on the Labour Government to support the banning of atomic weapons provoked a furious response. Will Arthur (South Wales) castigated the resolution as 'part of a policy that is world-wide in furthering the cause of a policy that is the opposite to that of a democratic Labour Government and of the [NUM], and is merely directed...to a continuation of a cold war that is undermining the forces of democracy and the forces that are attempting to preserve world peace.' Communist totalitarianism, Arthur concluded, was a greater threat to world peace than the atomic bomb and W.E. Jones claimed the resolution was directly inspired by the Conference of Culture for World Peace held by the Cominform in Poland (August 1948) and that it should be rejected in favour of supporting Bevin and the Labour government. The resolution was defeated by 421,000 to 261,000.[137] Lawther's assault on Communists in general and specifically on Paynter pulled no punches. 'Your economic outlook is as wrong as your political outlook', he fulminated, 'There are people who did their damnedest to prevent a Labour Party victory. There are people who would do their damnedest to wreck it by any means, and every opportunity is being used in order to do that.'[138] This assault produced uproar at the Conference.

The outbreak of the Korean War and the government's massive rearmament programme encouraged doubts about the reason for increased coal output. Bill Ellis (Northumberland) suggested the NEC tell the government 'Yes, we will produce the coal, but we will not produce the coal for war–war on the insistence of a foreign power or a capitalist power', and Jim Hammond warned 'we are not going to be able to electrify people about the production of coal to revive Britain when none of those who are calling out for coal are stating the real purpose for which it is required...none of them has publicly declared that the coal is required to pile up armaments in Britain.' A Scottish delegate challenged Watson's argument that extra coal would 'fill the baskets of our women...and put clothes on our backs and shoes on the feet of our children' and the miners had a right to say what should be done with the coal they produced.[139] This led to the familiar attack from Lawther who could not 'understand or appreciate the type of mind which eulogises the Stakhanovite in other nations, and sneers at the sloggers in his own pit' and compared the CP's attitude to the war in Korea to its advocacy of peace with Nazi Germany in 1940.[140] In an attempt to defuse conflict, Ernest Jones (the Vice President), on Sam Watson's suggestion, called for the remission of Resolution 35 (World Peace) on the understanding this implied nothing other than support for Labour's foreign policy. Compromise was prompted by the truce in Korea and Yorkshire, moving the resolution, agreed to remit. Jones described this as a 'gesture of goodwill' but warned of dire consequences 'if there is any possibility of any section of this Union, or any political organisation' departing from the compromise. Moffat agreed to remission even though the Scottish delegation was mandated to support the resolution, but Paynter's charge in the debate on the cost of living that inflation was the result of the rearmament programme provoked further uproar and broke the compromise. Glyn Williams, leading the South Wales delegation publicly dissociated the Area from Paynter and blamed Labour's rearmament programme on Soviet aggression. Cries of 'nonsense' were drowned out by a volley of interjections leading Jones to warn 'if there is another interruption the man responsible will go either out of *that* door or out of *that* one.' Williams pointed out – to cries of 'untrue' – that opponents of the Soviet regime ended up in Siberia.[141]

Abe Moffat, expressing the hope he would not be thrown out for supporting Paynter, argued 'the increase in the cost of living in this country was due to the domination of American Imperialism'.[142] Those opposed to Paynter argued that attacks on the Labour government ignored what would have happened under a Conservative government and 'if you have aggressors that give rise to Koreas, you are bound to have a speeding up of defence preparations...you are bound to have an increased demand for raw materials'. As there was no sign of the USSR disarming 'I would rather resist the possibility of attack on our freedom with some reduction in our standard life, than I would be a robot in a state controlled by dictators.' It was the Soviet threat which limited what the Labour government could achieve.[143] Renewed fighting in Korea led to Scotland seeking permission to move an Emergency Resolution condemning American bombing of power stations. This was ruled out of order by the Conference Business Committee (chaired by Sam Watson) and it needed a 75 per cent vote of Conference to over turn this decision. Blocking the resolution was interpreted as a reflection of increasing intolerance of dissent by a Business Committee determined to maintain a façade of unity in support of the Labour government. Lawther's rejoinder was that the decision was not an interference with democracy but the resolution should have been submitted earlier.[144]

What were the consequences of Cold War politics for the development of the NUM's politics? The first consequence was the ossification of the NUM's politics and the creation of a cohesive right-wing majority on the NEC. Moffat's description is extreme but does encapsulate the changed nature of the NUM's politics: 'The right-wing were not only anti-Communist or anti-Arthur Horner, they became in effect anti-miner, and against all strikes, no matter what circumstances led up to the strike.'[145] The strength of the right-wing on the National Executive was revealed when James Bowman resigned to join the NCB. The agreed right candidate for President, W.E. Jones, was nominated by 16 sections of the NUM whereas the left candidate, Abe Moffat was nominated by just Scotland and South Wales.[146] There were five members of the Communist Party on the NEC and they (and Horner as General Secretary) were muzzled by the requirement that NEC members must speak only in support of NEC decisions.[147] This imposition of orthodoxy was partly the result of the times: much was at stake – public ownership, the welfare state, full employment, even the survival of the Labour government – and in such a climate any threat to unity inevitably produced a hostile reaction. Especially important was the CP's changed attitude to production which invited open conflict because of the emphasis placed by the Labour government on maximum coal production.

The second consequence was to reinforce NEC dominance. The 1944 Rulebook intended the NEC would act as counter-weight to the Areas but the NEC soon began to assert itself as the representative of general interest. This can be seen in the NEC's control of the right to declare a strike official. Under Rule 41 the NEC could delegate the power to call a stoppage but an attempt by the Northumberland Area to use Rule 41 in a dispute at Barmoor colliery led to the NEC seeking legal clarification of Rule 41. The advice was that no cessation of work could take place without the *prior* approval of the NEC and there was no power of retrospective approval. If NEC approval was not forthcoming any strike called under Rule 41 was *de facto* and *de jure* unofficial and outside the protection of the NUM's rules.[148] The Barmoor decision was critical to the development of the NUM as it rendered official industrial action at the Area level impossible and put the decision to call *any* industrial action in the hands of the NEC. The NEC's ideological commitment to public ownership and the insistence that in a publicly owned industry there was no need to strike meant the NUM had effectively abandoned the right to strike.

A related factor which increased the influence of the NEC was the constructive interpretation of the vagueness of the 1944 Rulebook. Horner told the 1945 Conference that 'adjustments will have to be made in the light of experience of the working of the one organisation. If the organisation is to justify itself, it must bring greater unity among the coalfields, in addition to greater loyalty within the coalfields.'[149] These interpretations were all in the direction of increasing NEC influence. The political and economic pressures facing the coal industry, coupled with the need to work with the *National* Coal Board ensured that the *National* Executive was inevitably raised over the Areas. Some officials such as Glyn Williams feared over-centralisation and called for more, not less, sub-national democracy in the NUM especially now that key negotiations were taking place at national level. As opinion was strongly in favour of this Horner recognised the NEC would lose the vote but saw the resolution as retrograde as 'people are apparently still hanging on to old forms which grew up in entirely different circumstances and different situations from those which obtain at present.'[150] At the 1947 Conference the Rulebook was amended to permit more frequent Area Conferences but this proved largely symbolic and in 1949 an attempt to modify Rule 43 to ensure that proposed national agreements should be ratified by the branches was countered by a plea from Horner that 'we hope you will not categorically tie us down so that we cannot bargain even though it may be unpopular in a certain district, but beneficial to the membership as a whole.'[151]

The Cold War and the growth in the NEC's centrality dovetailed in a way which profoundly influenced the NUM's politics. Despite Gaitskell's prediction Horner chose the NUM:

> No one could accuse me of not being loyal to decisions once they had been made democratically either by the Executive or the delegate conference of the Union. My colleagues of the NUM...all knew that I took my instructions from the Executive of my Union and from no one else. No one suggested, or indeed could suggest, that my membership of the Communist Party meant that I went to my Party for instructions or consulted them on matters concerning the Union.[152]

Horner's problem was that this suggestion *was* made and in the climate of the late 1940s the charge carried considerable weight, which meant Horner trod a 'political tightrope' while his colleagues 'left [him] free to express his political views as an individual, but any attempt to use his office for political purposes brought immediate rebuke from his colleagues.'[153] The CGT controversy represented these limits, Horner's Communism kept him off the TUC General Council and out of the trade union movement's inner councils. By 1950 internal NUM politics were dominated by a determination that nothing should be permitted which might threaten or undermine public ownership and it made little difference whether this threat came from Communists or unofficial strikers. With Labour's loss of office in 1951 and the arrival of a Conservative government the nature of internal political conflict shifted to embrace a wider conflict about the extent to which the NUM should co-operate with a Conservative government.

Notes

[1] Cabinet Committee on the Socialisation of Industries (47) 23, 21 May 1947 Parliamentary Enquiries Concerning Nationalised Industries. Memorandum by the Minister of Fuel and Power. *CAB 134/688.*

[2] Cabinet Committee on the Socialisation of Industries 4th Meeting, 22 May 1947, item 4(ii) (I) p. 5. *CAB 134/688.*

[3] NUM Report of the Annual Conference, July 1947, p. 30. Hereafter, *NUM (ACR).*

[4] *NUM (ACR),* June 1945, p. 24.

5. D.N. Chester, *The Nationalisation of British Industry*, HMSO 1975, pp. 399-400 and W. Ashworth, *The History of the British Coal Industry. Volume 5 1946-1982: The Nationalised Industry*, Clarendon Press 1986, pp. 138-140.

6. *Joint National Negotiating Committee* (Workers' Side), 8 January and *NUM (EC)*, 10 January 1946. The NUM's members of the JNNC (Will Lawther, James Bowman, Arthur Horner, W.E. Jones, Abe Moffat, and Ebby Edwards) were the NUM's 'inner cabinet.'

7. Appendix I. Coal Industry Nationalisation Bill. Replies of the Minister of Fuel and Power to Points on which the National Union of Mineworkers sought Information, *NUM (EC)*, 14 January 1946. Ebby Edwards, the NUM General Secretary, was prevailed upon to join the NCB as Board member responsible for industrial relations and Arthur Horner was elected as his replacement. Horner was also asked to join the Board but he refused.

8. A.A. Berle and G.C. Means, *The Modern Corporation and Private Property*, New York, Macmillan 1947, p. 69.

9. *NUM (EC)*, 14 January, and Appendix I Letter Received from the Minister of Fuel and Power, 11 March, *NUM (EC)*, 14 March 1946. The clause required the NCB to recognise unions representing substantial parts of the workforce, not just the NUM. When the NUM split after 1984-85 this was used to abrogate the 1946 agreement recognising the NUM and include the UDM in the consultation/conciliation machinery.

10. NUM, *Report of the National Executive Committee*, May 1946, pp. 179-80, and *NUM (ACR)*, June 1946, p. 81.

11. *NUM (ACR)*, June 1946, p.16 and pp. 80-81.

12. NUM *Report of a Special Conference*, 20 December 1946, p. 17.

13. *NUM (EC)*, 19 December 1946, and F. Zweig, *The Men In The Pits*, Gollancz 1948, p. 160.

14. Socialisation of Industries Committee (47) 32, Taking Stock. Memorandum by the Lord President of the Council, para 8, p.2 and para 10, p.3. *CAB 134/688*. Hereafter, SI(M).

15. SI(M) (47) 43, 13 November 1947, Taking Stock. Memorandum by the Minister of Fuel and Power, para 34, p. 5. *CAB 134/688*. The most serious co-ordination problem was not between the NCB and Ministry of Fuel and Power but between the NCB and the Ministry of Supply which was responsible for the supply of mining equipment.

16. P. Williams (ed.), *The Diary of Hugh Gaitskell*, Jonathan Cape 1983, 18 June 1948, p. 72. Hereafter, *Gaitskell Diary*. Gaitskell was Shinwell's Parliamentary Secretary before taking over at the Ministry of Fuel and Power.

17. *NUM (EC)*, 8 January and 7 October 1948.

18. NUM *Report of the National Executive Committee*, May 1949, p. 226.

19. *Gaitskell Diary*, 12 July 1948, p. 75.

20. *Gaitskell Diary*, 12 August 1948, p. 80.

21. SI(M) (49) 33, Government Control Over Nationalised Industries. Memorandum by the Minister of Fuel and Power, 30 May 1949, para 7, p. 2. *CAB 134/690*.

22. SI(M) (49) 33, paras 14-15, pp. 3-4.

23. *NUM (ACR) 1947*, pp. 109-10.

24. *NUM (ACR) 1947*, p. 19 and F. Williams, *A Prime Minister Remembers*, Heinemann 1961, p. 93.

25. N. Fishman, 'Coal: Owned and Managed on Behalf of the People' in J. Fryth (ed.), *Labour's High Noon. The Government and the Economy, 1945-51*, Lawrence and Wishart 1993, p. 64.

26 Notes of a Meeting with the Minister of Fuel and Power, *NUM (EC)*, 17 August 1945. P. Slowe, *Manny Shinwell: An Authorised Biography*, Pluto Press 1993, p. 215.
27 NUM, *Report of a Special Conference*, 10-12 October 1945, p. 14.
28 *Report of a Special Conference*, p. 23.
29 *Report of a Special Conference*, p. 52.
30 Zweig, *The Men In The Pits*, p. 156.
31 Appendix B. Recruitment - Charter of Demands, *NUM (EC)*, 10 January 1946.
32 Appendix. Copy of a Letter Received from the Minister of Fuel and Power, 11 March 1946, *NUM (EC)*, 14 March 1946. Slowe, *Manny Shinwell*, p. 215.
33 NUM, *Report of the National Executive Committee*, May 1946, p. 286.
34 *NUM (EC)*, 8-9 October and 14 November 1946. The reason for this refusal was that the NRT's decisions were binding. So if it rejected the NUM's case that was the end of the matter and rejection might provoke serious unofficial unrest in the pits.
35 *NUM (ACR), 1946*, p. 69.
36 Appendix II. Notes of Meetings of JNNC (Workers' Side) and Discussions with Representatives of the NCB in Relation to a Five Day Week, 2-5 December 1946, *NUM (EC)*, 19 December 1946, and *Joint National Negotiating Council* (First Meeting), 17 December 1946. Hereafter, JNNC.
37 NUM, *Report of a Special Conference*, 20 December 1946, pp. 14-15.
38 K. Harris, *Attlee*, Weidenfeld & Nicoloson 1982, p. 33-4. Jay to Attlee, 2 April 1946 *PREM 8/440* is a good example of Jay's analysis.
39 5s *H.C. Debs*, 24 July 1946, cols. 68-70, for example. Slowe, *Manny Shinwell*, p. 217 and p. 219.
40 *The Times*, 25 October 1946. Shinwell's strategy is set out in CP (26) 232, 17 June 1946. *CAB 129/10*. Slowe, *Manny Shinwell*, pp. 222-23.
41 A.J. Robertson, *The Bleak Mid-Winter. Britain and the Fuel Crisis of 1947*, Manchester University Press 1987, p. 66.
42 Coal and Electricity. Memorandum by the Minister of Fuel and Power, 3 January 1947. *CAB 129/16*.
43 A. Horner, *Incorrigible Rebel*, MacGibbon & Kee 1960, p. 179.
44 For the management of the crisis see, Robertson, *The Bleak Mid-Winter*, pp.76ff. Attlee's committee's proceedings are in *PREM 8/443*.
45 *NUM (EC)*, 15-16 January, and Appendix X Statement on the Coal Crisis, in *NUM (EC)*, 13 February 1947. As well as calling for more effort from the mineworkers the NUM pressed for more mining machinery, improved transport to move coal from the pitheads, and better working conditions to attract labour into the mines. See Appendix IX. Note of Meeting between Representatives of the Union and the Committee set up by the Prime Minister, under his chairmanship, to deal with the Fuel Emergency, 27 February 1947, *NUM (EC)*, 13 March 1947. On 10 February Attlee took personal charge of the crisis and Shinwell was sidelined. In October 1947 he was replaced with Gaitskell.
46 A. Cairncross, ed, *The Robert Hall Diaries, 1947-1953*, Unwin Hyman 1989, 7 October 1947 (p.10).
47 NUM, *Report of a Special Conference*, 14 March 1947, pp. 24-27.
48 Report of a Meeting with the Prime Minister and Minister of Fuel and Power, *NUM (EC)*, 18 March 1947.
49 Appendix. Situation in the Coalmining Industry. Production Programme for 1947, 25 March 1947, *NUM (EC)*, 2 April 1947.
50 NUM, *Report of the National Executive Committee*, May 1947, p. 222.

51 Appendix I Meeting with the Prime Minister and other Ministers, 30 July 1947, *NUM (EC)*, 8 August 1947.
52 Meeting with the NCB, *NUM (EC)*, 12 August 1947.
53 NUM, *Report of a Special Conference*, 22 August 1947, p. 7.
54 *Report of a Special Conference*, pp. 9-18.
55 *Report of a Special Conference*, pp. 34-5.
56 Situation in the Industry, *NUM (EC)*, 10 September, and Meeting with the NCB, *NUM (EC)*, 17 September 1947.
57 NUM, *Report of a Special Conference*, 10 October 1947, p. 11.
58 National Coal Board, *Annual Report and Statement of Accounts for 1946-47*, para 58, p. 13.
59 Coal Production - Meeting with Minister of Fuel and Power and Members of the Coal Board, *NUM (EC)*, 7 October 1948. The NUM and NCB set up a Joint Production Committee which reported on 27 October but it offered no major departure from current policies. Its suggestions aimed at increasing attendance by involving the NUM in labour discipline as in wartime were rejected by a majority of NUM Areas.
60 National Consultative Committee, *Minutes of the 13th meeting*, 9 November 1948.
61 Extension of Hours Agreement, *NUM (EC)*, 20 February 1948.
62 Joint Committee on Production, Joint National Negotiating Committee, *Minutes of the 43rd meeting*, 12 January 1949. Hereafter, *JNNC*.
63 Meeting with the Minister, *NUM (EC)*, 12 April 1949.
64 Meeting with the Minister of Fuel and Power, *NUM (EC)*, 21 November; Situation in the Industry, *NUM (EC)*, 30 November 1950; and NCB, *Annual Reports and Accounts for 1950*, para 1, p. 1.
65 Meeting with the Prime Minister, *NUM (EC)*, 3 January 1951.
66 Meeting with the National Coal Board, *NUM (EC)*, 10 January 1951.
67 NUM, *Report of a Special Conference*, 17 January 1951, p. 9.
68 Meeting with the Minister of Fuel and Power, *NUM (EC)*, 28 March 1951.
69 NUM, *Report of a Special Conference*, 5 April 1951, p. 18.
70 Appendix I Joint NCB/NUM Committee on Production 2nd meeting, 8 March and 3rd meeting, 30 March 1951, *NUM (EC)*, 12 April 1951.
71 Coal Industry National Consultative Committee, *Minutes of the 34th Meeting*, 13 November 1951 hereafter, CINC), and Cairncross, *Robert Hall Diaries*, 24 October 1951, p. 174.
72 *NUM (ACR) 1952*, p.45, and *CINC* (39), 30 October 1952. See also, Cairncross, *The Robert Hall Diaries*, 28 May 1952, p. 228.
73 K. Middlemas, *Power, Competition and the State. Volume 1, Britain in Search of Balance, 1940-1961*, Macmillan 1986 discusses this problem at considerable length. C. Barnett, *The Lost Victory. British Dreams, British Realities 1945-1950*, Macmillan 1995, pp. 345-62 provides a powerful critique of the full employment commitment.
74 *JNNC*, (17) 22 October, (18) 5 November, (19) 13 November, (20) 14 November, and (21) 14 November 1947.
75 *NUM (EC)*, 18 September 1947, and *Report of a Special Conference*, 10 October 1947, pp. 28-29. The Special Conference voted by 71 to 39 to accept the wage offer of a national minima of £6 per week underground and £5.50 for surface workers and an increase of 3s 4d (15.8p) in all district daywage rates.
76 See Horner's introductory comments to the Special Conference on 20 November 1947.
77 Personal Incomes, Costs and Prices, *NUM (EC)*, 11-12 March, and 1 April 1948.
78 *NUM (ACR) 1948*, p. 64.
79 *NUM (ACR) 1948*, pp. 69-70.

Inclusion or Integration?   77

80 *NUM (ACR) 1948*, p. 82-84.
81 Coal Production. Meeting with the Minister of Fuel and Power and Members of the Coal Board, *NUM (EC)*, 7 October, and Wages, *NUM (EC)*, 28 October 1948.
82 National Reference Tribunal 19th Award, *JNNC* (46), 4 May 1949. NCB resistance can also be seen in the conflict over concessionary coal allowances in Lancashire and Cumberland.
83 NUM, *Report of the NEC*, May 1949, p. 226.
84 *NUM (ACR) 1949*, pp. 184-86. A South Walian delegate blamed the NUM's failure not on complacency but 'because of our loyalty to the Government' (p. 187).
85 *NUM (ACR) 1949*, pp. 191-92.
86 *NUM (ACR) 1949*, pp. 193-94 and pp. 198-99 for Horner's speech.
87 *NUM (EC)*, 16 November, and 15 December 1949. Lawther and Bowman were members of both the TUC General Council and the Special Economic Committee.
88 Trade Union Wage Policy, *NUM (EC)*, 28 December 1949.
89 NUM, *Report of a Special Conference*, 29 December 1949, pp. 10-11.
90 *Report of a Special Conference*, p. 18.
91 *Report of a Special Conference*, pp. 22-4.
92 *Report of a Special Conference*, p. 32.
93 *Report of a Special Conference*, p.35.
94 *Report of a Special Conference*, p.22.
95 *Report of a Special Conference*, p. 29.
96 *NUM (ACR) 1950*, p. 92.
97 *NUM (ACR) 1950*, pp. 93-4. Lawther ruled calls for the NUM merely to *consider* using its industrial strength out of order.
98 *JNNC* (54), 21 June 1950. At a subsequent meeting the NCB stressed the burden of its accumulated debt (£42.5m) and the current surplus was be used to create a Reserve Fund. These profits came from exports not the home market. *JNNC* (55), 22 June 1950.
99 *JNNC* (57), 11 October and (59), 19 October 1950.
100 NUM, *Report of a Special Conference*, 2 November 1950, p. 9 and p. 11.
101 *Report of a Special Conference*, p. 18.
102 *Report of a Special Conference*, pp.20-21.
103 *Report of a Special Conference*, pp. 23-4.
104 Meeting with Ministers, *NUM (EC)*, 3 January 1951.
105 J. Davis Smith, *The Attlee and Churchill Administrations and Industrial Unrest 1945-1955: A Study in Consensus*, Pinter 1990, pp. 96-7. *NUM (ACR) 1945*, p. 81. Only Durham opposed the resolution. In May 1946 the Midlands Area strongly opposed CP affiliation to the Labour Party. *NUM (EC)*, 1 May 1946.
106 Fishman, 'Coal: Owned and Managed on Behalf of the People', p. 73. See also P. Weiler, *British Labour and the Cold War*, Stanford University Press 1988.
107 *NUM (ACR) 1947*, pp. 15-17.
108 See *The Times*, 9 February 1948 for a discussion of Communist influence in various trade unions including the NUM.
109 Press Statement, *NUM (EC)*, 29 May 1947.
110 General Secretary, *NUM (EC)*, 8 January, and Labour Party Conference, *NUM (EC)*, 12 February 1948.
111 *Gaitskell Diary*, 6 January 1948, p. 51. Gaitskell wrongly believed Horner would side with the CP.
112 *Gaitskell Diary*, 8 October, p. 87, and Coal Production - Meeting with the Minister of Fuel and Power and Members of the Coal Board, *NUM (EC)*, 7 October 1948.

[113] Appendix I. Speech of Mr A.L. Horner to the Conference of the French General Confederation of Labour held in Paris from 11-15 October 1948, in *Report of the Special Sub-Committee of the National Executive Committee*, 11 December 1948, p. 18. Hereafter, Special Sub-Committee. The Sub-Committee (appointed by the NEC on 28 October) was composed of Sam Watson, W.E. Jones, Abe Moffat, Will Arthur, and Joe Kitts. All, except for Moffat, were orthodox Labour supporters.

[114] The French trade union movement had split between the socialist Force Ouvriere (FO) and the Communist Confédération Général du Travail (CGT). The miners, affiliated to the CGT, struck over pay and conditions but the strike was widely interpreted as a Communist inspired attempt to destabilise the government, from which the French Communists had withdrawn in June 1948, on the orders of the Cominform. See V.R. Lorwin, *The French Labor Movement*, Harvard University Press 1954, ch. 8 and pp. 129-131 for the miners' strike.

[115] International - Mr Horner's French Visit, *NUM (EC)*, 28 October 1948. A letter of protest from the Scottish Area was ignored.

[116] Horner's full case is set out in Appendix II of the Sub-Committee report.

[117] *Sub-Committee Report*, para 39, pp. 9-10.

[118] The CP's line was that Marshall Aid was intended to revive West German economic and military power as a precursor to war with the USSR. Disrupting production was, therefore, a means of preserving peace.

[119] *Sub-Committee Report*, para 68, p.15. The Foreign Office Information Research Department fed anti-Communist information to sympathetic trade union leaders and Lawther used this material in newspaper articles. Weiler, *British Labour and the Cold War*, pp. 207-208.

[120] Moffat, *My Life with the Miners*, pp. 269-70, and *NUM (EC)*, 16 December 1948.

[121] Trade Unions and Communism, *NUM (EC)*, 17 November, and 16 December 1948. *Defend Democracy* was made up of two General Council statements, *Warning to Trade Unionists* (27 October) and *Communist Activities Examined* (24 November).

[122] Horner, *Incorrigible Rebel*, p. 189. Lawther and other leading trade unionists were also wined and dined by the US ambassador. R. Pearce ed., *Patrick Gordon Walker. Political Diaries 1932-1971* (London, The Historian's Press 1991), 29 May 1948, p. 177. Also present were Arthur Deakin, Vincent Tewson, Hilary Marquand, Michael Foot, Fred Peart and Jim Callaghan.

[123] B. Pimlott, ed, *The Political Diary of Hugh Dalton 1918-1940, 1945-1960*, Jonathan Cape 1986, 27 July, p. 399 and 28 July 1947, p. 401. Stephenson was director of British Security Co-ordination in the Western Hemisphere (1940-46) and Donovan had undertaken special missions for Roosevelt in Western Europe (1940-41) and became Director of the Office of Strategic Services (1942-45).

[124] *Gaitskell Diary*, 22 October 1947, p. 40.

[125] Cairncross, *Robert Hall Diaries*, 18 September 1947, p. 6.

[126] Weiler, *British Labour and the Cold War*, p. 268.

[127] Davis Smith, *The Attlee and Churchill Administrations and Industrial Unrest*, p. 99. For an example of these attacks see *The Daily Mirror*, 8 January 1949 and *NUM (EC)*, 13 January 1949. *JNNC* (42), 12 January 1949. See also Weiler, *British Labour and the Cold War*, p. 215.

[128] Weiler, *British Labour and the Cold War*, p. 255 and p. 353.

[129] *NUM (ACR) 1949*, pp. 20-22.

[130] *NUM (EC)*, 10 February 1949. In May the NUM resolved not to send delegates to the 1949 WFTU Congress in Milan. The World Federation of Trade Unions had been set up in 1945 as a common organisation for all national trade union movements but the

Inclusion or Integration? 79

changing policy of the USSR led to the American, British, and Dutch unions to set up the International Confederation of Free Trade Unions.

[131] Paynter was the left's standard bearer at this time and was NUM General Secretary between 1959 and 1969. His memoirs are uninformative on NUM politics in this period. W. Paynter, *My Generation*, George Allen & Unwin 1972.
[132] *NUM (ACR) 1949*, p. 106
[133] *NUM (ACR) 1949*, p. 108.
[134] *NUM (ACR) 1949*, p. 109.
[135] *NUM (ACR) 1950*, pp. 17-20.
[136] *NUM (ACR) 1949*, pp. 67-8.
[137] *NUM (ACR) 1949*, pp. 129-33.
[138] NUM, *Report of a Special Conference*, 2 November 1950, p. 21.
[139] Extension of Working Hours, NUM, *Report of a Special Conference*, 5 April 1951, p.17, p. 22 and p. 24. This last delegate also quoted from Lawther's 1947 President's speech which stated the miners would not produce coal for a war with the USSR.
[140] *NUM (ACR) 1951*, pp. 23-24.
[141] *NUM (ACR) 1951*, pp. 181-82.
[142] *NUM (ACR) 1951*, p. 186.
[143] *NUM (ACR) 1951*, pp.191-92.
[144] *NUM (ACR) 1951*, pp. 129-30.
[145] Moffat, *My Life with the Miners*, p. 268.
[146] *NUM (EC)*, 9 February 1950.
[147] NEC members who disagreed with official policy developed means of demonstrating their opposition to a policy which they were compelled to support and they could (apart from Horner) use their Area base to articulate an alternative policy.
[148] Barmoor Colliery, *NUM (EC)*, 1 June 1945.
[149] *NUM (ACR) 1945*, p. 32.
[150] *NUM (ACR) 1947*, p. 40. The resolution was passed by 339,000 to 278,000 but had no visible effect on policy.
[151] *NUM (ACR) 1949*, p. 42. Conference agreed to remit the resolution to the NEC.
[152] Horner, *Incorrigible Rebel*, p. 190.
[153] Horner's obituary, *The Times*, 5 September 1968.

Chapter 3

The Politics of State Capitalism

Introduction

During the nationalisation bill's passage Harold Macmillan goaded the Labour benches:

> This Bill vests the ownership of all the colliery undertakings in a board of nine men – nine men not elected by, or even containing a single elected representative of, the mining community. It is not nationalisation in the old sense of the word...This is not Socialism; it is State capitalism. There is not too much participation by the mineworkers in the affairs of the industry; there is far too little. There is not too much syndicalism; there is none at all...To the men, the new owners will mean the Board. However gifted or eminent they may be, they will be more remote and more soulless than the old owners.[1]

Macmillan's analysis soon struck a chord with many mineworkers. This chapter examines three aspects of the nationalised coal industry. The first section examines the consultation system and argues that it should be seen as a type of sectoral corporatism whose purpose was to prevent conflict in the coal industry spilling out into the wider political system by enmeshing the workforce on a framework of rules and procedures which emphasised an ideology of cooperation. The second section explores the NCB's managerial ethos and questions the extent to which this differed significantly from private ownership. Of course, there were major improvements and significant differences between the two forms of ownership but the overall situation remained one of mineworker and union subordination. The third section explores the nature of government control over the price of coal. That the NCB was clearly not a private sector company is shown by its failure to exploit its market power and that it was forbidden by law to make profits. These differences are of secondary importance as the NCB's purpose was to support the wider rate of profit by subsidising industry's energy costs.

The Politics of Consultation

The political dimension to the industry's affairs resulting from public ownership raises the question of the extent to which the NCB's social relations of production differed fundamentally from those of a private company. Concentration on legal forms of property ownership ignores the social relations of property and does not distinguish between nominal and effective 'ownership', the latter being concerned with controlling the corporation's operations and legal ownership matters far less than effective control. Nationalisation was a change of legal ownership but it was less clear that it represented a significant shift in the social relations of production.

Unofficial strikes and absenteeism grew markedly in the war and the Reid Report had warned radical measures were needed to combat this if the industry was to extract the full benefits of modernisation. Reid's 'social contract' had called for a drastic restructuring of workplace relations and a dramatic shift in the unions' role from defending the workforce against management to becoming an agent of workforce control, or at least managers of discontent. As in all contracts there were penalties for non-compliance, 'the mineworker has certain...fundamental duties, and failure to carry them out destroys, to the extent of his failure, *his claims to the rights we have enumerated.*'[2] The NEC agreed that any mineworker who refused to accept the NUM branch committee's ruling in any pit dispute 'would mean that the Union would refuse to intervene in the event of disciplinary action being taken by the management.'[3] Some colliery managers went further and called for 'the imposition of sanctions for dealing with absenteeism and indiscipline' and that these be coupled to the implementation of the Miners' Charter. In Lancashire, for example, the Production Officer reported 'the state of relations was very bad. There was a mood of indifference and the men are apathetic. This is reflected in the number of local disputes and unofficial stoppages'. His Kent colleague believed exhortations to produce more coal were 'futile' as the men would not listen and the Scottish Production Officer believed any campaign to boost output ought to be delayed 'until the Government makes some definite statement upon their attitude' to the Charter as the miners would not respond 'until they can see some definite advantage is to be gained from their efforts.'[4]

Public ownership of the mines *per se* was expected to produce a marked improvement in performance and that the conciliation machinery would be taken as proof that there were no longer disputes in the industry, only differences, where the NUM and NCB sought to settle grievances 'without resort to strike action or to court proceedings.' The NEC warned 'that the action of a small minority in the industry in participating in unofficial stoppages should be strongly deprecated.'[5] Public ownership, a Labour government, and a new enlightened employer obviated the need for industrial action. The NUM leadership argued forcefully that the nation had finally recognised its obligation to the miners, the miners must now recognise theirs to the nation. The NUM would naturally continue to defend its members but 'we also have a right to expect that there shall be an end to unofficial strikes...it is a crime against our own people that unofficial strikes should take place...No stoppage can be justified, having regard to the present dire need for coal.'[6] Despite 'the greatest single concession in our history' far too much output was being lost and Horner described the behaviour of voluntary absentees and unofficial strikers as 'intolerable' and that this 'minority must be regarded as an alien force, and treated as an enemy of the true interests of the majority of the miners of this country.'[7]

The complexity of the industry's political and industrial relations situation was demonstrated by a major unofficial dispute at Grimethorpe colliery in Yorkshire.[8] The strike, which began on 11 August and ended on 15 September 1947, was caused by the implementation of the Five Day Week Agreement. At Grimethorpe faceworkers traditionally left the pit when their 'stint' (the amount of coal they were contracted to produce in a shift) was completed but the Agreement led to an increase in stints of 2 feet which resulted in 2,600 mineworkers striking at Grimethorpe. Ten days later the first Yorkshire pits came out in sympathy and on 26 August the Divisional Chairman, Major-General. 'Mickie' Holmes, announced all strikers were deemed to have broken their contracts and were liable for prosecution. This resulted in the strike spreading further.[9] Despite rescinding this threat one third of Yorkshire's pits were on strike by 8 September, the numbers striking were inflated by the understandable attractions of a warm summer and racing at Doncaster. In total, 63 pits were affected and 594,300 tons of coal were lost. The proposals which had sparked the strike had been negotiated by the Pit Production Committee but rejected by the NUM branch, it was then passed to the Disputes Committee for resolution and the strikers were urged to return to work by the Yorkshire NUM, the NUM, the TUC, the NCB, Labour MPs and Emmanuel Shinwell while it came to a judgement. They refused. Grimethorpe encapsulated many domestic and foreign doubts about nationalisation: 'Seldom has a strike – even in the United States – been watched with such apprehension. Grimethorpe is becoming a household word and symbol of the decay which is supposed to surround the British coal industry.'[10] Grimethorpe was emblematic of the attitudes and behaviour that public ownership was supposed to eradicate but equally, it was not typical of the pattern of post-nationalisation industrial conflict, even in the strike prone Yorkshire coalfield, which was overwhelmingly characterised by large numbers of small, localised brief, but highly disruptive, stoppages.

The NCB advised the Divisional NCBs that faced by such disruption they had two options: institute legal proceedings for breach of contract or close those collieries with a poor performance and disciplinary record except where a colliery was a major long-term contributor to total output. Hyndley reminded the NUM that their President and General Secretary had affirmed frequently their support for management taking whatever steps were needed to eradicate indiscipline. In November the JNNC was informed that 1.5m tons of coal had been lost in disputes and the Board contrasted this with the millions being spent on 'improving the wages and conditions of the miners and referred to the irresponsible attitude of the men.' The NEC could do nothing other than concede the force of the Board's arguments.[11]

Attempts to make the NUM responsible for disciplining the workforce, as was advocated by the Joint Committee on Production, failed. Its report was rejected outright by the Area unions who refused to become management enforcers. When pressed by Ebby Edwards on what the NUM intended to do to stop absenteeism and disputes the NUM was at a loss and they were told further concessions depended on 'a substantial improvement...in attendance and effort resulting in increased output and reduced costs.'[12] A detailed analysis of stoppages condemned many of the causes as trivial and concluded 'there was...no excuse for stoppages.' This criticism caused some frustration amongst NUM leaders who believed they were doing everything humanly possible to end disruption and the union was cooperating fully with the NCB at all levels. Hyndley did not dispute this but was concerned that most stoppages were over issues which could be easily resolved at pit level. To Hyndley's consternation Lawther suggested 'it was unlikely that the conciliation Machinery would be fully recognised and become a familiar and automatic feature of mining life until a new generation of miners had grown up who were thoroughly acclimatised to its use.' Another difficulty was that 'the prestige of the Conciliation Machinery with the men varied according to its rewards.

When it gave judgements in favour of the men it was naturally popular, when it awarded against them it became equally unpopular.'[13]

These problems were illustrated by a dispute in the Lancashire coalfield. The NCB sought assurances from the NUM that the union was neither overtly nor covertly supporting the strike and was doing everything in its power to secure a resumption of work. These assurances were given and the Lancashire Area was instructed to abide by the Conciliation Agreement and help prevent the disputes spreading to the Yorkshire and Durham coalfields otherwise 'the consequences would be incalculable.' The Board remained concerned because even when the strikers returned to work 'no attempt had been made to use the conciliation machinery' and it expressed its outrage 'that there had been repeated threats and warnings of further strikes unless some concessions were made.' A statement by the local NUM which purported to support the men was condemned by the NCB; 'Did the Union attach no value at all to the National machinery for conciliation which they had jointly agreed, to the sanctity of national agreements, to the maintenance of their own authority as a national body?'[14] The dispute grumbled on until the Autumn of 1949 and the Lancashire Area withdrew from the joint production committee, refused to operate the Extension of Hours agreement and even sought to withdraw safety men from the pits.[15] The dispute was settled in October on terms favourable to the NCB.

Hyndley described these attitudes as 'most disappointing' and he renewed his attack on the widespread refusal to use the conciliation scheme. Of the 1,500 disputes in 1949, 80 per cent were classed as minor and only 16 per cent were referred upwards to pit committees or beyond. Disputes were concentrated in three coalfields – Yorkshire, Scotland and South Wales – and some pits had clearly developed a habit of indiscipline and conflict. Both the NUM and NCB agreed there was no justification for this level of unrest.[16] Iestyn Williams, an NCB member, complained of 'So many stoppages in an industry where the conciliation machinery was generally considered to be second to none, pointed to a failure on the part of the men to understand their responsibilities under the agreements between the NUM and the Board. In some Areas there were now more disputes than before Vesting Day.' Williams argued it was the Union's task to persuade their men to use the machinery but Ernest Jones disputed the imputation 'that the men were always at fault.'[17] A common complaint in the NUM was that the men were accused of being largely responsible for the disputes; 'the implication seemed to be that it was always the men who were to blame for [any] failure to use the conciliation machinery.

Managers...sometimes failed to use the machinery.' This charge prompted Ebby Edwards to ask the NUM for details of management failures 'But no single case had been brought to the Board's notice since the Vesting Date.'[18] The causes of these disputes and their remedy was a frequent item on CINC agendas but seldom did the discussions produce any significant proposals.

Ministers saw worker involvement as essential if there was to be cultural revolution in the coal industry and Morrison urged his colleagues to recognise that consultation and conciliation procedures were of 'the highest importance' and should be used 'with imagination and enterprise.' Drawing on the NCB's experience he argued that nationalised industries should be a model for the rest of industry and 'a worker should feel that there is an avenue open to him by which he may make his views and ideas known to the management of the industry.'[19] The consultation procedures were deliberately separated from the conciliation (collective bargaining) machinery which was designed to provide a conflict resolution mechanism, whereas the former was designed to integrate the workforce into the structure and ethos of public ownership and thereby encourage a shift from a culture of conflict to one of co-operation. Not surprisingly 'the consultative' as it was invariably called, was portrayed as a major shift in the industry's power structure.[20] However, the 1947 Conference debate on the Pit Conciliation Scheme broadened dramatically into a critique of the developing NCB power structure.

Willie Pearson declared that whilst he and the Scottish Area opposed unofficial strikes delegates ought to understand the frustrations of the workforce. One of these was that the General Managers and representatives from the Divisional NCB on the Divisional Disputes Committee 'are not playing the part they ought to be playing.' Pearson believed that 'the division between us when it comes to disputes is as sharp today as it was with the coal owners' as NCB representatives went to the DDC to fight grievances rather than resolve them. He claimed 'the result of the present administration will be to destroy the psychological effects of nationalisation' and there had already occurred a sharp deterioration in the mood in the pits after the initial euphoria. 'Our people', a delegate claimed, 'are by the hundreds expressing the opinion that the old procedure, the old approach, is being made, and is being made in a stronger fashion than ever hitherto.'[21] Abe Moffat argued the NUM had to accept its obligations and take more responsibility for production issues. Consultative structures were, however, excluding the mineworkers from discussions over the industry's technical capacity which was seen as the sole province of the mining engineer. Reconstruction would bear down hard on the mineworkers and if they were to bear these costs they should be involved in the key discussions. Joe Hall (Yorkshire) warned that unless the union was involved in these discussions it would be very difficult to secure the men's co-operation in reconstruction.[22]

War, the creation of the NUM in 1944 and public ownership in 1947 progressively located parochial issues in a wider national context. After 1945 the NUM had to come to terms with a much more complex industrial and political environment. In his study of the early post-war mineworker Zweig found:

> two basic conceptions still struggling for supremacy within the unions. One is put forward mostly by the younger and more militant members, those mainly responsible for the lightning strikes, the other by older, more moderate members, who preach the doctrine of self-discipline and responsibility with the recent accession to power [of Labour] and nationalisation.[23]

Branch officials were traditionally perceived not as workforce representatives but as delegates whose job was to defend their members. Public ownership was intended to root out this adversarialism but there was a general recognition this would take time and mechanisms would have to be provided to enable management and union to co-operate in the industry's interests whilst fulfilling their traditional roles. Gradually a co-operative ethos would supplant the tradition of conflict.

In 1942 the Ministry of Fuel and Power issued a circular which provided the primary objective of industrial relations in the coal industry; 'The first aim is to prevent a stoppage of work, or to get work re-started if a stoppage has occurred, pending an attempt at a satisfactory settlement being reached on the issue involved through the proper negotiating machinery.'[24] The aim was to bring the two sides together to negotiate and implement their own settlement, a philosophy based on the historical experience of government being dragged into the contorted politics of the coal industry. The political theory and practice of coal nationalisation required the maintenance of a balance between the centrifugal and centripetal forces at work in the industry. There was an ever present tendency towards conflict and unrest in the mines which given the centrality of coal in the post-war political economy could not be permitted to spill out of the pits. The structures, procedures and ideologies balancing these centrifugal tendencies were considerable and included the consultation and conciliation system, the NUM's Rulebook, loyalty to the

Labour government, gratitude for public ownership, and fear of the Conservatives.

One observer of the industry wrote, 'The miners' unions have now reached the highest point in their status and power. Is there any force which could resist them, and is there anything which could be denied to them if they really make up their mind to struggle for it?'[25] Central to public ownership was the mineworkers effectively abandoning the right strike. The NUM decided to trade power for status. The NUM achieved its status because of coal's importance, manpower budgeting, and the total collapse of the legitimacy of private ownership as a result of the 1942 crisis. By 1946 the NUM subscribed to a 'social contract' of output in return for public ownership which would deliver decent wages and conditions, so creating a virtuous circle. To demonstrate that the NCB would be different the NEC endeavoured 'to test out the character of the new employer...we were informed that the [NCB] was to function as a good employer, and we have been wondering what significance we should attach to that word "good".' Horner argued the acceptance of the Miners' Charter and the negotiation of the Five Day Week at a time of severe economic crisis proved the NCB was different to the coalowners, whilst the consultative machinery ('which will control the destinies of the industry') gave the mineworkers an unprecedented voice in the industry's affairs. These concessions had been won despite the scepticism of the NCB and government and represented a 'terrific gamble.' Horner reiterated 'we have placed all our dependence on our own people and on our faith that they will respond...in a manner that will shock and surprise the population of this country.'[26] The country's need for coal, reinforced by the co-operative ethos of public ownership ruled out, so far as the NUM leadership was concerned, industrial action and this had obvious implications:

> Everybody should know, in our position of buying and selling labour when mining labour is so scarce, if we had chosen to command the moon, we could have got it or the country would have been unable to carry on. Do you think we do not know our strength? Or that we were foolish enough not to believe in our power? We know our power, and if we have refrained from using it, it is because of our concern for the affairs of the country, for we are part of our country, and we cannot remain prosperous and be surrounded by destitution in other industries.[27]

The NUM did not exploit their market power and as a symbol of the new order had agreed to work through agreed rules and joint procedures to secure the mineworkers' best long-term interests.

Neither war nor public ownership eradicated strikes which were blamed by union leaders, managers and government on workforce indiscipline and, with very little evidence, on 'unconstitutional elements'. After 1940 the NUM had become central to enforcing labour discipline in the pits and keeping costs down which were key purposes of corporatist-type politics, and public ownership did not alter this. Ironically, whilst the NUM enjoyed greatly increased access to management, the structures and procedures which provided this access served to limit the NUM's influence. This could only have changed if the NUM had been willing to confront both the NCB and government using methods forbidden in the new order. This reticence, the demand for coal and the managerialist-efficiency ethos which permeated CINA ensured that managerial authority rapidly asserted itself, aided by the NUM's insistence there be clearly defined 'sides' in the industry. Despite the wartime dominance of the MLNS (as a result of manpower budgeting and the physical and political weight of Ernest Bevin) coal remained semi-independent. This reflected the long established general understanding that the Mines Department (hitherto part of the Board of Trade) was the coal industry's sponsor and was responsible for handling disputes in the industry. This was continued by the Ministry of Fuel and Power and Sir Frank Tribe lamented the 'long tradition of what might be termed isolationism and on both sides one finds a desire that the mining industry should be regarded as something separate and distinct from other types of industry.'[28] This isolationism was not broken down by the war and it was then enshrined by public ownership.

In the negotiations over the CINA Shinwell identified the NUM and the NCB as the key bargaining partners and agreed that average wages in mining should not be below the industrial average.[29] The National Conciliation Scheme and the Pit Conciliation Scheme, the eight Divisional Labour Directors and a National Board member (Ebby Edwards) responsible for industrial relations obviated the need for government involvement in the industry's internal affairs. As the Ministry of Labour told its Regional Industrial Relations Officers in January 1947 'when disputes or apprehended disputes are reported, [action] should be confined to ensuring that both parties are aware of the existence of the trouble and that the negotiating machinery is, in fact, working.'[30] The conciliation machinery was not intended to abolish conflict as this was impossible, its purpose was to channel conflict into a system of rules and consensus building reinforced by a 'neutral' arbiter, the NRT, whose decisions were binding. If the NUM and NCB could not reach agreement, the dispute had to be referred to the NRT for arbitration and both sides were committed to accepting its decision. As NCB finances deteriorated, and wage inflation became the dominant concern of economic policy, the NRT (which did take into account the Board's balance sheet) became politicised in the minds of many of the NUM's members, leading to demands that the NUM withdraw from the conciliation scheme. The implications of this were obvious and the suggestion was resisted stubbornly by the NUM's leadership using all the resources available to them under the 1944 Rulebook. What was not clear, however, was the extent to which this system could contain sustained pressure from within the coalfields.

The purpose of the consultation machinery was to minimise conflict and maximise consensus in the industry over reconstruction which could not be achieved painlessly. This required the NUM advance the industry's interest over those of individual groups of members. The closure of Waleswood Colliery in North Derbyshire testified to tensions inherent in the structure of power in the nationalised coal industry and symbolised the growing discontent in the industry. This had to be resolved by the fullest consultation between the NUM and NCB, especially as there was a fear that the situation might be exploited by 'organisations not connected with the Union.'[31] In early 1948 the NCB decided that Waleswood be closed and its reserves mined from neighbouring collieries. This provoked a stay-down strike which enjoyed the tacit support of the Derbyshire Area (NUM). Hyndley, recognising the importance of Waleswood for the industry's future as reconstruction would inevitably lead to many more closures, sought the NUM's advice. Lawther recommended management 'prevent food being sent down the pit' whilst Horner pleaded the matter be left to the NUM and that no legal action be considered. The National Consultative Committee agreed to allow the NUM 'take the necessary steps to deal with this matter.'[32] Waleswood became the paradigm for pit closures (other than those caused by exhaustion or geological problems) as it was based on an 'assessment of technical and economic factors' which was 'the responsibility of the Board alone' which would 'do everything in their power to minimise hardship to the workmen concerned.' This required the co-operation of the workforce and union 'to secure to the country the full benefit which the scheme is designed to achieve.' Hyndley accepted the workforce must be consulted but the management of the industry was, and would remain, the Board's responsibility. The NUM's role was to secure worker acceptance of Board decisions taken in the interests of the industry and the country.[33]

The issues raised by Waleswood were discussed at the 1948 NUM Conference. Horner's presentation of the NEC's position stressed the political delicacy of the times. The NCB's first Annual Report was about to be published (leaks pointed to a substantial loss) and that as a result of reconstruction 'hundreds of pits...are now in their death throes, and are due to die.' The NUM had to decide in favour of 'modern mines with the latest up-to-date machinery [or] depend on the slave driving of physical labour' which would reduce the miners' standard of living and undermine the Labour government. 'Old pits have got to die. The Union's business is to see that the social consequences...are carried through as humanely as possible', Horner concluded.[34] Delegates argued closures were not the issue, but the NUM's exclusion from key decision making. Dai Llewellyn (Somerset) complained that 'we are not being informed of reorganisation plans....we are never informed of what is taking place on the technical side.' Bert Wynn (Derbyshire) saw this omission as the heart of the Waleswood controversy. The stay-down strike resulted from the miners not being informed about the wider issues so what they saw 'were their brethren...being attacked.' If pits were going to close in this way conflict was inevitable.[35]

Public ownership placed the NUM (at all levels) in an ambiguous position because although management and union remained clearly demarcated, public ownership imposed (as NUM leaders pointed out constantly) wider responsibilities which might generate internal conflict. Indeed, the consultative system could itself generate unrest because 'In times of crisis, when unpleasant decisions had to be faced, an irresponsible member of a Colliery Consultative Committee could be the local and temporary hero of a pit just by opposing the Union lead and telling the miners what they wanted to hear. Membership of the Union did not always imply readiness to accept Union disciplines'.[36] Inside the Cabinet Morrison wondered whether management 'did not need stimulating to take more energetic action' on worker involvement. Gaitskell's response was that NUM-NCB relations were excellent 'the real problem arose at the rank and file level. The difficulty was to get the miner to take enough interest...in view of the long history of bad relations in the industry he doubted whether it would be wise to try and rush the water.'[37] Balancing 'union' and 'pit' politics became an increasingly complex activity for the NEC. The 'Ashton' study conducted in the mid-1950s in the Yorkshire coalfield laid great emphasis on the increased complexity of post-1947 production politics. The authors of the study argued:

> nationalisation destroyed the unambiguous and simple position of branch officials and substituted for it one which is far more complex. While the possibilities of achieving the branch's aims are increased during the new dispensation, so are the occasions for misunderstandings between the men and their representatives. Simple opposition is easily understood and a certain glamour adheres to it. Co-operation is always more complicated and it is not easy to reconcile the miners to the idea of co-operation with the management.[38]

Coal Is Our Life presents a simplistic picture of pre-nationalisation industrial relations where there is ample evidence of bargaining and negotiation between management and union in often very unpropitious circumstances.[39] Management union co-operation existed before public ownership and it provoked disquiet from the membership, but there is no doubt there was a perception that the gap between union and men was growing. Mass Observation's study of Thorne and Rossington contains the following exchange,

Collier: My dad used to talk a lot about the unions. He used to say they were all for the men.
Investigator: Aren't they now?
Collier: Yes, they are, but now the Coal Board acts like a gaffer for the Unions.

A second collier commented on the NUM and its officials; 'They're hand in hand with them. They're not working for [the men] any more. It doesn't matter what grievances the men have they call it unofficial...'[40]

The NUM and NCB regarded the consultative machinery as the key to the future and Zweig described it as 'the most controversial subject' in the new industry. The reaction of a Nottinghamshire (praised by Reid for its co-operative culture) manager to the JCC was typical; 'The workers are not, in fact, interested in management, and they do not care for discussion on the intricate problems of production. We have no single improvement or betterment of any kind out of these Committees.' Consultation confused traditional roles in the industry, meetings were often concerned with parochial issues, and used to air grievances rather than solve output problems, and were therefore distrusted by many managers. A further difficulty was persuading mineworkers to accept JCC decisions as the union members were perceived as having 'gone over' to management.[41]

JCCs were perceived by the mineworkers as symbolic participation, ratifying decisions taken by pit managers or higher up in the NCB's bureaucracy. Mass Observation found a strong sense of exclusion among the workforce; 'Little appears to be known about the work of the consultative committees, and feeling toward the Committee is somewhat synical [sic] and despondent.' The mineworkers expected that the committees will be 'over-ruled by the gaffers' or take too long to come to a decision.[42] The mineworkers' sense of isolation demonstrated that 'under nationalisation the miner is not being given enough say in the running of the industry' and 'this adds up to a widespread feeling, none the less real for being expressed in differing and vague terms, that the thinking miner expected a new status, a new responsibility to come to him immediately'.[43] The sense of not being consulted undermined the system which, in turn, encouraged a low level of interest in the industry's wider problems and led to mineworkers concentrating on the immediate work situation. The consultative system was trying to reconcile the irreconcilable, the demands of production versus those of welfare, and not surprisingly neither management nor unions felt easy with the new system which became dominated by bargaining not collaboration, reinforcing the ever-present tendency to conflict as well as laying the basis for an élite consensus embracing management and union.

The NCB's Ethos and the Politics of Cultural Change

When the nationalisation bill was published in August 1946 the NEC minuted 'that this is but the first step towards the ideal for which we have striven for so long' and that the further development of the industry depended on the 'cooperation on the part of all employed therein.' The mineworkers were urged to 'recognise the necessity of breaking with the past' in order to ensure the success of the industry.[44] Breaking with the past proved difficult when so little seemed to change. Mineworkers accepted the need for 'Bosses' but many believed the wrong kind of boss had been appointed; "I'll raise my hand to the Labour government. They've nationalised the mines, just as they promised. Now, they want to change the Coal Board and they'll make a clean sweep', and, "The Coal Board...is just the old with no change. I don't say that nationalisation hasn't its faults but it's a great improvement".[45] These comments reflect much of the ambivalence to public ownership in the pits.

The narrow social and occupational background of many of those appointed to the NCB and the continuation of the existing management strata produced inevitable accusations of a lack of sympathy towards nationalisation, and sometimes charges of actual sabotage by managers wedded to the old regime motivated by a determination to undermine the industry's success and workforce morale. MO found a powerful perception in the pits that managers as a group were hostile to public ownership and their resentment led them to treat the workforce badly and express disdain for the consultative machinery.[46] Attlee articulated the government's dilemma with clarity, agreeing some managers did not relish change but 'You couldn't suddenly create a whole lot of new mining engineers, for example. You couldn't sack all the old mine managers and put in people who knew nothing about it. There was a good deal of difficulty here and there, I've no doubt, because the older dogs couldn't learn new tricks.'[47] Gaitskell deployed a classic Burnhamite view of the technical/managerial expert's motivations arguing that:

> the vast majority of the miners realise that the idea of sweeping away all the old managements and managers overnight was really quite ridiculous...a lot of managers were opposed to nationalisation...That itself does not mean, I suggest, that they are not prepared to do their jobs properly under nationalisation...the overwhelming majority of the managers have accepted the fact that nationalisation has come to stay...and will be completely loyal to the industry, and that they are more and more coming to realise the technical possibilities which nationalisation has opened up.[48]

There was a great deal of truth in this view but it is equally true that there was a great deal of suspicion amongst mangers.

In his memoirs Jim Bullock, a Yorkshire colliery manager who was sympathetic to public ownership, gives a picture of the resentment of many managers at the new regime. He writes, 'Overnight management status changed, we were no longer king of our village, the number of officials above the Colliery Manager grew and grew until it was often referred to as the 'Heinz Soup set-up'.[49] Though sympathetic, Bullock was adamant that managers were no longer able to run 'their' pits in the way they thought best, negotiations were taken out of their hands, their decision-making freedom was reduced and there was an increase in form filling. Despite his technocratic sympathies Gaitskell was forced to admit that 'There are too many people in management positions who are still strongly opposed to nationalisation and quite willing to make trouble with the Coal Board and the NUM. The Coal Board has not yet won their loyalty, and it cannot do so easily and at the same time retain the goodwill and support of the Union.'⁵⁰ Many mineworkers were convinced that public ownership would only work if managers were 're-educated': 'Nationalisation is a great thing. But the government isn't running the pits, the Coal Board is running the pits, and sits too much under the old régime.'⁵¹

The Fabian Report *Miners and the Board* suggested one possible solution to this problem; 'If all the coalowners – or a sizeable number of them – had been shot or otherwise liquidated, and if the leading managers and technicians had fled in terror to the United States or South America, the miners would undoubtedly have felt an immediate consciousness that the pits were their responsibility and theirs alone no one can say what the effect would have been.'⁵² A Stalinist solution was out of the question but it expresses the scale of the problem. At the 1949 NUM Conference Durham sponsored a resolution calling for a meritocratic appointments procedure. The delegate moving the resolution (which was carried unanimously) argued restructuring the industry depended not on management dominance but on workforce engagement. The perception 'that we are not yet capable of administrative posts' and management's dominance of appointments caused resentment and frustration and 'We will not get a new spirit into the industry by those methods.'⁵³ The resolution had no effect on NUM policy.

Central to Morrison's conception of the publicly owned industry was the technocrat appointed because of his proficiency. Appointments made on political, ideological or social grounds 'would involve deliberately depriving ourselves of many able people as well as exposing the government to severe criticism...we must avoid a system of patronage in the public service.'⁵⁴ Public ownership would work only if the workforce had confidence in its managers and the Fabian Society described their survey results as 'disquieting.' The miners had little knowledge of the NCB and what they had was often derived from gossip and the press; there were felt to be too many survivors from the old regime who were unsympathetic, too few real experts, and too many ex-generals. Significantly, there was no strong conviction that the miners should be represented directly on the Board.⁵⁵ Public ownership was grafted onto a workforce accustomed to mistrusting all managers.

Nationalisation's success depended on a cultural revolution but the industry's historical legacy was so dire, and so great were the immediate pressures on the miners to produce the coal and participate fully in the industry's affairs, that the inevitable tensions threatened to overwhelm the industry before it stabilised. Doubts about the willingness and ability of the NCB to bring about a cultural revolution were voiced before Vesting Day. In December 1946 South Wales, protesting about managerial appointments, asked the NEC 'to reconsider its role [sic] of non-intervention' in NCB appointments. The NEC refused and reaffirmed its policy of not accepting any responsibility for NCB appointments.[56] South Wales complained that the NCB's appointments policy was 'tending to destroy the confidence of the miner in the new Administration.' Six of the eight nominations for General and Production Managers in South Wales were ex-Powell Duffryn managers and 'we have a General from India who is absolutely ignorant of mining conditions. We have had in South Wales other men who were ignorant of mining. Our last Controller was a Judge who had spent most of his life in India...Now we have got another General from India.' In the Anthracite District one Regional NCB chairman was accused of having appeared on a platform with Oswald Mosley. Appointments from the private ownership era undermined nationalisation's credibility as, for example, '[Powell Duffryn] were considered to be the most efficient company in South Wales...but their efficiency was only possible on the basis of the damnedest tyranny that any company has ever exercised.' Appeals for greater output from managers such as these would be ignored 'because though we appreciate the fact there is a change in ownership...there is the same team with the same jerseys on.' The North Western Divisional Production Officer, an ex-employee of Manchester Collieries Ltd, was described as an excellent mining engineer but his attitude had provoked great hostility in the past and he had compounded this by appointing Manchester Collieries personnel to key positions who were known to be responsible for poor industrial relations.[57] 'It is true', Gaistskell told the Socialisation of Industries Committee, 'that some may have been opposed to nationalisation but most were not very politically minded and are deeply interested in the success of the industry in which they work; in any case, being human and knowing that their careers lie in the industry, they are anxious individually to make a success of their jobs.'[58] Horner conceded there were problems with some managers but the nationalised industry had only been in existence for a few months and he pleaded for patience and pragmatism as 'This is the first time we have ever travelled this way. It is all experimental...We are at present unduly dependent upon the technicians from the old regime until we discover a new attitude towards technique and begin to build our own forces to take the places of the people from the old regime.'[59] Public ownership was the *first* step towards socialisation which would create both a co-operative ethos and a new managerial class drawn from the workforce, meanwhile the mineworkers were urged to give the industry time and maintain discipline.

Ironically many mineworkers agreed with management's complaints about the diminution of their authority: 'Miners' feelings on this point are especially evident in the stress laid upon the impersonality and unnecessary difficulties created by "remote control". The NCB was not an everyday topic of conversation in the pits but there was a deep-seated feeling the Board was not on the miners' side and that its bureaucracy (extending to the NUM) meant decisions could not be taken on the spot so exacerbating tension needlessly. This problem was related to the perceived lack of co-operation and enthusiasm for public ownership by pit managers which, in turn, reinforced the traditional suspicion of management.[60] The counter argument was that the NUM was now in an incomparably more powerful structural position than under private ownership and 'if we can swallow the [NCB], then we can swallow the gnat which actually is the District Coal Board.' A new order was emerging and these 'hangovers' from private ownership 'know their world is ending', they would have to adapt or go. Mineworkers 'have to become so politically conscious that we can divide the dross from the gold' and deal with 'sabotage....in the higher realms of the coal industry.' The NEC conceded there were complaints from every coalfield but the NUM had a clear choice between maintaining its independence or becoming involved in appointments and thereby assuming managerial responsibilities. They chose the former. Furthermore, there was a wider political issue at stake because 'if we go out campaigning against people appointed, we could create such a state of dissatisfaction, dispute and unrest in the minds of our people that we might unwittingly sabotage the success of nationalisation.' Bowman agreed managers would have to learn that times had changed, public ownership had removed the profit motive and increased the power of the NUM which would neutralise the effects the NCB's appointments.[61]

Reliance on existing management was justified to the NUM on three grounds: 'because we need them; because it would be grossly unfair to attack them...and because they are not wreckers and saboteurs.'[62] The root of the problem was the shortage of skilled management personnel because 'There are few men in the industry with experience of planning big reconstructions because there was little of this done between the wars.'[63] Members of the Board had considerable experience in the coal industry or in other large organisations but no one anywhere in British industry had experience of running anything so large and complex as the NCB. Dependence on existing managers was reinforced by the low cultural level of workforce as 'until there has been more experience by the workers of the managerial side of industry, I think it would be almost impossible to have worker-controlled industry in Britain, even if it was on the whole desirable.'[64] The industry's problems required a high level of leadership from management which, in turn, required a clear administrative structure to handle the production crisis and begin the industry's transformation.[65] Jim Bullock recalled 'We had no people with the necessary skills, qualifications and/or experience who could tackle the tremendous task of organising – under one vast undertaking – a thousand different pits formerly owned by six hundred different companies.'[66] Ministers conceded there was a tension between centralisation and decentralisation in the coal industry. The former achieved economies of scale and uniform administration, the latter encouraged initiative and it was noted with approval that large private companies favoured decentralisation 'because of the growing complexities of administration and co-ordination.'[67] The problem was that the NUM was deeply suspicious of decentralisation.

Attlee was pragmatic, 'Well, one had to put in people who understood [management] techniques. It's no good thinking a lot of amateurs can run a complicated business' and his Minister of Fuel and Power warned colleagues that 'it took a long time to secure really efficient management after nationalisation.'[68] Joe Gormley, the NUM President between 1971 and 1981, scarcely mentions nationalisation in his memoirs because 'the fact was that it made precious little difference to us at the time...The deputies stayed the same. The managers stayed the same. And we never saw the owners anyway so a change of ownership couldn't have much effect on our daily lives.'[69] The face of the new order stimulated problems and doubts:

> Now we started getting real shocks. Men we'd never heard of were appointed to the biggest jobs, men who incidentally, had never heard of – never mind knew – anything about the Coal Industry. Gossip started as to how they had got these positions...We had admirals, generals, retired colonels, a discarded Food Controller, landed aristocrats, the lot.[70]

Whilst one can detect here the eternal complaint of the 'practical man' at externally imposed change, Bullock's comments expresses a widely held perception at the time of nationalisation. The miner's perception of a burgeoning and highly paid bureaucracy feeding off his labour was based on his immediate work situation where there was an increase in the number of technical and supervisory grades many of whom were perceived as lacking practical mining experience and some were appointed in dubious ways. A South Wales miner wrote 'between the men at the pits and the Coal Board it seems to him that a new 'snooper' class has jumped into authority, and flaunting that new power and authority more arrogantly than before in his experience.'[71] Ninety five per cent of the Fabian Society's sample agreed bureaucracy had grown after 1947 and believed many appointments were unnecessary or were appointed to snoop on the workforce. These were often described as 'spivs.'[72] The NCB in London was perceived to be so distant as to be outside the mineworker's consciousness whereas Area and Divisional officials were seen as a drag on the industry and a burden on producers.[73]

Ministers saw ex-admirals and generals as the only source of personnel accustomed to managing large scale operations, as well as being imbued with the ethos of public service. Gaitskell had been very impressed by Rear-Admiral Woodhouse, the chairman of the Kent Divisional Board, and believed 'There is no doubt that there are advantages in having men from the Services in these jobs for the simple reason that doing something for its own sake and because the job comes naturally. They have been brought up to believe in service and so, as it were, take to nationalisation.'[74] Gaitskell wanted a 'really national figure' such as Mounbatten or Field Marshal Montgomery to replace Hyndley (who was ill), someone 'with a good sense of public relations and a capacity for making people work as a team under him.' Gaitskell's enthusiasm cooled after he met Montgomery at a Buckingham Palace dinner. They got off to a bad start when Monty told Gaitskell he had advised one of his ex-generals who had resigned from the South Wales board to denounce the NCB. Gaitskell concluded Monty's 'extreme egotism, lack of humour and being a bore' made him unsuited to be Hyndley's successor.[75]

The NCB's critics received a major fillip with the resignation of Sir Charles Reid on 13 May 1948. He followed up his resignation with series of articles in *The Times* which were reprinted as a pamphlet which argued the industry's structure was fatally flawed. The divisional NCBs and management in general were, Reid wrote, 'deprived of initiative and responsibility' becoming the passive recipients of instructions from London where 'there is no one in complete control of the executive.' Pit managers did 'not feel themselves to be in charge of operations as they were under private enterprise. They are unable to make decisions which ought to be within their province, and therefore their status and authority in the eyes of the staff and workmen are reduced.' Reid urged the NCB be decentralised and restructured into 'manageable entities' which was so important 'in an industry where personal energy and enterprise are supremely important.'[76] Gaitskell acknowledged the force of Reid's criticisms. On 27 April he had met Hyndley and the Board to discuss Reid's criticisms and resignation threat. In response Hyndley set up a Policy and General Purposes Committee and sought external specialists to recommend changes in the NCB's structure. Gaitskell invited S.R. Burrows (ex-chair of Manchester Collieries and the LMS railway) to become a part-time Board member. Burrows was 'an old Tory but just the right type that we want, because he knows the industry and is a business man, not a technician; also a man of considerable administrative experience on a large scale.' The appointment of part-time board members was strongly opposed by Ebby Edwards as they were outsiders but significantly Edwards was overruled.[77]

Attlee regretted the lack of trade union involvement in the management of the boards. When asked if more trade unionists should have been appointed he replied; 'We used what we could get. They weren't always willing to cross over, nor were their men always willing for them to go in: a curious contradiction, because they talked of Labour running the show and yet when you put a trade unionist to help run a nationalised industry they tended to regard him as a bosses' man.'[78] Even when NUM officials were willing to join the NCB they were often not the first choice. Shinwell, for example, wanted Arthur Horner but he refused because 'I might have reached the point when as a member of the Board I had to do something which I disbelieved, and I would then have been obliged to throw up the job or break my heart.'[79] The NEC recommended two of its members for appointment to the NCB but Shinwell rejected them, arguing they were more suitable for the regional boards. Shinwell 'desire[d] that the person appointed...should occupy the most prominent position, which would be indicative of the importance which the Union attached to the work of the Board.' Abe Moffat reports being asked by Shinwell to become NCB Labour Director, Sam Watson was also approached but neither he nor Moffat would permit their names to go forward, as a result Ebby Edwards was put under enormous pressure to join the NCB.[80] An informal approach had been made to Edwards who, with the endorsement of the NEC, was willing to serve.[81] Hyndley complained that 'getting the right man' at Divisional level, 'was one of my greatest problems, and no one caused me a greater headache than your own Union, because in every Division I went to I could not get the men I wanted.'[82] This was the context to James Bowman's appointment as Chairman of the Northumberland and Cumberland Divisional NCB in late 1949. Appointed with the full backing of the NEC, James Bowman was the first NUM official to become a divisional chairman, leading Lawther to comment, 'there has been a bit to say about some of the Chairmen now and again, and therefore you cannot object to the "brass hats" appointed from the other side and not accept the appointment of someone from our side.'[83]

One year after Vesting Day B.L. Coombes identified five benefits the mineworkers had gained from public ownership: security of work and wages, increased holidays, better pithead facilities, and the Five Day Week. Despite short-term problems Coombes argued 'we must agree that the Coal Board has tried to be a good employer. One year is not enough to undo the mismanagement and neglect of a quarter of a century, nor is it enough to dispel the hatred and distrust that neglect has left in the miners minds.'[84] Three years later Mass Observation found that:

> Unqualified approval was given by just over three in five people, qualified approval came from one person in seven, and those who disapproved, numbered only just one person in ten. The reasons most commonly given for disapproval...were that the old owners of the mines still wielded considerable power, and were sometimes using that power to sabotage the present effort, that nationalisation of the industry had created un-necessary jobs and something of a muddle generally.[85]

Few mineworkers seriously questioned either the principle of public ownership or its positive effects but when attitudes were probed MO found a number of deeply rooted criticisms. Whether these criticisms were accurate was of secondary importance to the fact that the mineworkers perceived 'defects in organisation and mis-handling of particular situations.'[86] One of public ownership's justifications was that it would end the trench warfare between management and men. Between 1947 and 1959 strikes in mining accounted for 70 per cent of all strikes, 36 per cent of workers involved and 19 per cent of working days lost in an industry which employed 2 per cent of the workforce. The typical strike was brief (less than three days) and involved relatively few workers (but could affect many more) costing about 1.2m tons of coal (0.6 per cent of saleable output).[87] The cultural shift from conflict to cooperation leading to an upsurge in output did not happen and the margin between having sufficient coal and a fuel crisis remained narrow, the tonnage lost due to industrial disruption and absenteeism were politically significant.[88] There were no official national disputes until 1971 but industrial relations in the 1940s and 1950s were worse than under private ownership.[89] The tonnage lost due to industrial disputes peaked in 1956 (2.164m tons), the largest number of strikes was in 1958 (2,224), the largest number of workers involved was in 1947 (308,000), and the largest number of working days lost (1,112,000) was 1955. After 1957 there was a marked reduction in the level of unrest in mining as a result of the industry's decline.[90] In 1951 disputes over pay accounted for 53 per cent of disputes (60 per cent if dissatisfaction with allowances and bonuses are included), the second largest cause was conflict over methods of working and colliery organisation (20 per cent) and so 'the main source of discontent in this respect was associated with fluctuations in earnings produced by changes in seam conditions, as well as by men having to transfer between seams of different paying capacities and managers withholding allowances for abnormal conditions as a form of disciplinary sanction.'[91] Wages were a complex and volatile confection of national rates, piece and day wage rates and allowances, a structure guaranteed to cause the maximum amount of unrest. The introduction of new working practices as the industry was reorganised and mechanised could cause major disruption as happened with the 1947 Five Day Week which cost 838,000 tons (50.6 per cent of all output lost due to industrial action) and changes in concessionary fuel allowances in Lancashire and Cumberland in 1949 which cost 383,000 tons (24.8 per cent of total losses).

In the first ten years of public ownership disputes in mining accounted for an average of 76.6 per cent of all stoppages in British industry. Coal was undoubtedly the most strike prone industry but aggregate figures conceal a geographical distribution of unrest which was concentrated in three NCB Divisions, the Scottish, the North Eastern (the Yorkshire coalfield) and the South Western (primarily South Wales).[92] These Divisions accounted for 54 per cent of the NCB's pits and 50 per cent of its wage earners but 80 per cent of the industry's stoppages and restrictions. More typical was the speedy settlement of disputes. In 1952, for example, of 11,666 pit level disputes, 11,100 (95 per cent) were settled at pit level, only 566 (5 per cent) were referred to the official disputes machinery or required the services of an umpire. Of the 19 cases referred to the Divisional level, 15 (79 per cent) were settled by the District Conciliation boards and 4 (21 per cent) by District Referees. There were no marked regional variations as about 90 per cent of all disputes in every NCB Division were resolved at pit level except for the West Midlands where the settlement rate was about 75 per cent.[93]

Commentators saw the historical and psychological legacy of private ownership as the most important obstacle facing the post-war coal industry. Zweig found 'the past is deeply ingrained in their minds. Whenever you start a conversation with the miner on the pits, he invariably begins by telling you about the [1926] Coal Strike.'[94] The central theme of *Coal Is Our Life* is the *reinforcement* of old attitudes and suspicions by public ownership. The mineworkers had made material gains but as late as the mid-1950s Ashton had (for example) no pit-head baths, no canteen just a tea hut open twice a day, and poor underground lighting. Despite public ownership the mineworkers did not experience a fundamental shift in their structural position in the industry because; first, 'the actual changes have been absorbed into the miners' traditional ideology rather than transformed it. Secondly, changes within the mining industry, and the quantitative improvement of the miners' position...have been unaccompanied by any profound modifications' in the mineworkers overall subordinate position in the industry's power structure.[95] The mineworkers' self-image contained a deeply ingrained belief in their isolation from the rest of society which lay at the heart of their defensiveness; this 'is the logical result of the miners' history' and public ownership was not going to transform these attitudes overnight and even though the mineworkers saw public ownership 'as the all embracing solution to the problems of the industry. Today, while there is every indication that the principle of nationalisation is thoroughly approved of...the Coal Board, visible sign and symbol of authority, is attracting to itself a complex of "anti-attitudes".[96] In 1948 the NEC complained about Mass Observation and the Fabian Society enquiring into the Board and stirring up resentment but felt compelled to conduct a similar review in response to growing disquiet in the coalfields.[97]

Presenting the NEC's Report on discontent in the pits to the 1948 Conference, the discussion took place in the presence of Lord Hyndley and Sir Arthur Street, Horner identified five main issues. First, excessive London-based bureaucracy out of touch with local conditions. What mattered most, Horner argued, was not the source of decisions but their consequences; 'I do not know of a single instruction intended to bring about a deterioration in the conditions of a single one of our members.' In any case, the NEC preferred a strong centre rather than see a return to the anarchy of district autonomy. Second, Horner denied managers had lost power and authority and any manager who claimed this, he argued, was not fulfilling the 'responsibilities of his job.' The manager should be in charge, but his authority must be based on consultation rather than autocracy or dictation and Horner agreed there was no reason why pit committees should always be chaired by the management side. Third, whilst many mineworkers had criticised the growth of superfluous appointments few concrete examples had been brought to the NUM's attention. In fact, there had been virtually no increase in the numbers in traditional managerial grades. New technical grades had been created and Horner confessed he was 'staggered' by the criticism; 'Safety Officers: we have always wanted greater supervision by men free from any other responsibility...Who asked for Training Officers? Who wanted to bring them in?...We did'. These appointments were the fruit of public ownership and NUM pressure, the hang-overs from the old days would be got rid of as soon as possible. The most serious area of concern was the conflict between Production Officers and Labour Officers (usually ex-miners). The ethos of public ownership implied a balance between the two but the NEC's evidence was that 'Production representatives at policy level are exercising a far greater influence in the determination of questions than the Labour representatives are doing.' Labour was the largest single cost in the industry and Production Officers were inevitably drawn into wages and conditions questions which encouraged the Production side 'to endeavour to dominate Labour representation within the Board's administration.' Labour Officers were too few, enjoyed a lower status than mining engineers, and were isolated from key decision arenas even though public ownership was intended to raise labour's status and influence over the industry's administration. Without this there would be no real cultural change producing disillusionment which meant 'the time may be coming when we shall have to consider direct representation of the Union in the Coal Board administration.'[98]

A Yorkshire faceworker (to the fury of his Area President) responded with an impassioned accusation that 'the Executive and our head officials, and most of the Coal Board, have lost touch with working conditions.' Mineworkers were subject to the same pressures from managers ('who are the same today as we worked under for the old coalowners') to produce more coal and cut costs, pressures which were not felt by the NUM Executive who ignored their members disquiet. Managers paid no attention to the NUM because the NUM had given up the right to strike and had agreed to compulsory arbitration. 'Conciliation is all right', he concluded, 'if you have both got a whip, but if they have taken your whip away from you, while can still use theirs, then it becomes a travesty'.[99] Abe Moffat argued that suspicion of the Board's composition did not exist at District and National level where there was a strong mutual confidence but 'what we have not succeeded in doing, is to create confidence in the minds of the miners of this country.' Without this confidence the future of public ownership was imperilled and Moffat argued confidence had to be based on effective worker participation and 'That seems to me the big problem, not the criticising of whether or not we have a Training Officer here or a chauffeur there.' Moffat was convinced that the Production/Labour officer conflict showed the need to create a new cadre of administrators sympathetic to the wider aims of public ownership. This was taken up by Jim Hammond who contended 'that the actual running of this industry is in the hands of experts of a past regime...supported by people from outside the industry...They cannot act differently from the environment from which they have come.' Hammond concluded, 'There has got to be more people from our own ranks going into that machinery...Is it impossible for people like ourselves to administer this industry?'[100] This was a discussion of an NEC report, as no resolution was put no vote was taken.

The problems of the state industries were taken up by the TUC which circulated a list of questions about employee participation to affiliates.[101] Ministers were worried by charges of increased, unresponsive and superfluous bureaucracy but Gaitskell took comfort from the fact that '[M]ost of the criticisms are...general and not specific in character which suggests that they are not well founded.'[102] The NUM's analysis acknowledged the consultative machinery had 'shortcomings', the most serious of which were 'that no functional change has taken place between management and workers', and 'advice proffered by the workers' side is insufficiently regarded.' Consultative Committee agendas were often determined by management and union representatives were denied the technical information necessary to participate fully. Little effort was being put into training union officials and workers for technical and management roles and the NEC was told 'there is far too much fishing in universities and public schools for trainees for managerial posts, which may eventually lead to class distinction and conflict.' Promotion to managerial posts was thought to be too dependent on 'the good-will of the Area Manager or Sub-Manager, and that this oftentimes depends not upon ability but upon favouritism.' The Report's concluding comment was pessimistic:

> After more than two years of nationalisation, whilst relationships have to some extent improved, there remains a good deal of dissatisfaction. The good will that has been built up so laboriously since Vesting Day is being replaced by cynicism. The Union recognises that whilst it is the primary duty of a trade union to deal with matters affecting wages and conditions of its members, in a nationalised industry and under conditions of planned economy it must use its maximum effort in the spheres of technical and productive activity and finance, especially when the level of costs or production are so decisively determining the remuneration which is to be made to those employed in the industry.[103]

This private appraisal contrasted strongly with the NUM's optimistic public position.

NCB appointments policy continued to cause disquiet. In the West Midlands Division, for example, there had been a reduction in the number of union appointments and an increase in part-time members without consultation of the NUM. The NEC was asked to consider reversing its policy of non-involvement in the NCB appointment's process and that any NUM member taking an NCB post be allowed to retain his union membership. The NCB apologised profusely for not consulting the NUM but insisted it remain solely responsible for appointments. Hyndley wanted more officials from the workforce 'but on several occasions when he had approached the Union representatives, his overtures had been met with refusal.'[104]

At the 1948 NUM Conference no fewer than five resolutions were composited calling for an inquiry into the NCB and more time was allocated to this issue than any other at the Conference. Tommy Degnan (Yorkshire), for example, argued public ownership could not deliver real change until 'the composition of the Coal Board is altered and that trade-unionists or people with trade union experience who have a socialist outlook and a socialist background should form the majority on the [NCB] and on the Divisional Coal Boards.' Abe Moffat insisted investigations were a diversion 'and we should be determined to convert nationalisation into socialisation by adopting...the principle of workers' participation at all levels in the nationalised industries.' Horner acknowledged the force of the criticisms and alluded to a growing suspicion that some managers were seeking to revive private enterprise practices. However, he also counselled delegates to remember the enormous gains made since 1945 and warned 'we intend to hold on to nationalisation of the mining industry whatever it means. Whether it be a Labour or a Tory Government they shall not reinstitute private enterprise in this industry except over the dead bodies of the manpower in the industry.' The NEC accepted the resolution on condition this should not be interpreted as criticism of public ownership or the NCB, but as part of an attempt to improve public ownership.[105]

Criticism of the NCB soon took on a ritual aspect. There quickly emerged a widespread acceptance in the NUM that there would be no fundamental power shift within the nationalised industry but that there had been significant material improvements. At the 1950 NUM Conference, for example, a resolution expressing alarm at the growth of the NCB bureaucracy received only cursory attention. Despite three-and-a-half years of public ownership 'the feeling still persists in the pits that we have handed the control of the industry back to the people we bought the pits from' and the NUM was warned 'the miners in the pits are beginning to look upon the Coal Board in many respects in the same way as they looked upon the old coal owners'. Lawther responded by repeating Horner's 1949 analysis and warning that criticism of the NCB was playing the 'reactionaries' game (a General Election was due), nevertheless he announced the NEC would hold another enquiry.[106] The enquiry had a low a priority as the NEC were anxious to avoid political embarrassment even though the NEC had been denied information about the NCB's proposed development programme, *Plan For Coal*. This was vital to the industry's future yet the NUM's members on CINC protested at the NCB's failure to consult.[107] The NEC's enquiry team first met in March 1951 and decided to investigate the South Western Division but it proceeded slowly because of the pressure of other work.[108] Subsequently, the Sub-Committee decided to visit every coalfield but this would take time, further delaying the report. This decision was, in part, prompted by the return of a Conservative government in 1951 which made the Sub-Committee's work politically sensitive and it did not wish to give the opponents of public ownership any justification for any intervention. Many of the NUM's internal criticisms of the NCB could be found in the Conservative Party's critique of the nationalised industries. This changed the Sub-Committee's purpose as 'we want to establish the clear fact that these industries are a great success, and in no industry has nationalisation succeeded to a greater degree than in the coalmining industry.'[109]

The 1951 NUM Conference criticised the appointment of part-time Board members and the conduct of labour relations by the NCB's Labour Department. These resolutions attracted little attention, they were formally seconded and no delegate expressed a wish to speak. W.E. Jones (NUM Vice President) spoke on the 1949 and 1950 resolution and the work of the Special Sub-Committee. Jones conceded that it had taken two years for the Sub-Committee to begin work but asked delegates to remember the scale of the task. Information was only just becoming available and he expressed the hope that a report could be presented to the 1952 Conference. More seriously Resolution 36 (Scotland) sought to remove Clause 9 of the Conciliation Agreement whereby unresolved national questions were referred automatically to the NRT, to permit the reference of a disputed claim to the membership before it was submitted to the NRT for arbitration. In effect, the resolution was calling for an end to compulsory arbitration, so raising the prospect of strike ballots. The resolution was based on growing hostility to the conciliation machinery and a sense that it was operating against the NUM's interests and was biased towards management. Those opposed argued approving Resolution 36 would signal that the NUM had conceded public ownership had failed to create a new culture in mining and approving it 'means going back to the old days of struggle and strife.' Horner agreed the conciliation machinery was not perfect and the NUM did not always get what it wanted 'but because we are aggrieved about a certain situation do we throw away a machine that has given us so much?'[110] The resolution was easily defeated.

The Price of Coal

About 60 per cent of the NCB's total production costs were wages so cost control meant wage control but wage control implied conflict with the NUM.[111] Coal's central role in the economy and the NUM's potential power transformed the price of coal into a major government concern. The NCB's commercial and financial obligations were to make 'supplies of coal available [in] such quantities and at such prices, as may seem to be best calculated to further the public interest in all respects, including the avoidance of any undue or unreasonable preferences of advantage' (CINA S.1(I) c) and 'the revenues of the Board shall not be less than sufficient for meeting all their outgoings properly chargeable to revenue account on an average of good and bad years' (CINA S.4 (I) c). Coal was no longer to be mined for profit and customers would not pay the full marginal costs of coal, but the average cost of good and bad seams. This presented the industry as one cost unit permitting cross-subsidisation and the NUM concluded this principle ought to apply to wages. However, sustaining this required ever higher productivity increases in the most profitable coalfields.

A Cabinet Office appraisal of 1956 reviewed the complex effects of coal price increases. Postponing a proposed increase as it was unlikely that 'the economy will be able to stand the shock of deferred increases in rail transport charge and coal prices came almost simultaneously.' Allowing wage increases to filter through into price increases might be 'sound economics' but would have potentially serious political and economic consequences as 'simply to raise the price of coal may well result in the worst of both worlds. The unions will use it as an argument that as [price] stabilisation is not succeeding they must re-insure themselves by further wage pressure during the year.' A refinement might be to approve the increase 'but to combine it with a movement of calculated aggression on the wage/price front.' This would involve a declaration that pay increases will not be awarded unless accompanied by greater productivity but this would infuriate the NUM and given the existence of full employment some employer was bound to give way.[112] During the 1950s both the NCB and government were anxious to reduce the industry's accumulated deficit but they faced constant upward cost pressure from wages, safety and welfare measures, reconstruction, raw materials, and coal imports which the government charged to the Board. CINA made the NCB responsible for covering its costs on an average of good and bad years but the NCB only had bad years which inevitably meant an increased role for ministers in the industry's pricing policy.

Ministerial control of coal prices had 'an appreciable and lasting effect' on the NCB's commercial activities because it prevented the Board building cash reserves by charging a market price for coal.[113] At the time of the Munich Crisis in 1938 the owners and the government concluded a 'Gentlemen's Agreement' that coal prices would not be increased without ministerial approval. This was extended voluntarily in 1946 by the NCB but Hyndley made it clear that the Board 'would, at some stage wish a firm indication from Ministers whether the industry was to be run for the public benefit or for the benefit of those working in the industry. There was a clash of interests here which would need to come to a head sooner or later.'[114] Interfering with the NCB's pricing policy was always going to be more of a problem for the Conservatives than for Labour. Soon after coming into office the Conservative government was faced by the need to raise pit-head prices by 10 per cent to meet increased wages and raw material costs. Ministers felt they had no choice other than to accede, especially as the NCB was running a deficit. Government policy was not to permit the creation of a substantial surplus as 'there was a psychological advantage, which the [NCB] themselves appreciated, in working with a small deficit' as a substantial surplus would encourage further wage demands from the NUM.[115] Continuing losses meant that coal prices, and therefore the extent of ministerial involvement in the industry's affairs, became a perennial item on the Cabinet agenda in the 1950s.

At the end of 1953 the NCB and government agreed the minister would be informed 'in good time' of any proposed price changes and the Board would 'pay regard to my judgement of where the national interest should be deemed to lie in relation to such proposals.' In the event of a disagreement, or if the minister overruled the NCB, the Board could request a written explanation from the minister.[116] The minister had no *statutory* power to control coal prices and the Coal Board asked neither to be released from the Gentlemen's Agreement, nor did it test its legal validity. This was because the NCB accepted the legitimacy of ministerial influence and a ministerial refusal to increase prices was a valuable negotiating tool vis-à-vis the NUM. The Gentlemen's Agreement was a major restriction on the right to manage, damaged the industry's finances and added substance to the NUM's charge that coal was subsidising manufacturing industry. Of the ten applications made by the NCB for an increase in coal prices, four were granted but for less than the amount requested, one was refused, and five were implemented after a delay. Horner frequently reminded the NUM 'when we are dealing with the [NCB], we are in effect dealing indirectly, and sometimes directly, with the Government of the day, whatever its colour may be.' The NUM was convinced that improved pay required increased coal prices but coal prices 'are of major political as well as economic importance.'[117] J.R.A. Machen (Yorkshire) insisted 'We cannot allow this industry to be a milch cow for the others, unless it is recognised that the first charge is a reasonable wage for our men, and in that relationship we have to proceed to get what we consider to be the reasonable and adequate standards that our men require.'[118]

The 1951 and 1953 price increases had no discernible effect on the NCB's deficit. Moreover, the NCB wished to make greater use of differential pricing, charging higher prices for coal types that were in high demand (6.5 per cent for large coal, 5 per cent for coking coal and 3 per cent for other grades) and this would require Cabinet approval. The NCB's deficit projected for 1954 was £30m and as the productivity increase had been allocated to improve the pay of the daywagemen so a price rise was inevitable. 'I think it would be unwise', Geoffrey Lloyd, the Minister of Fuel and Power, warned Cabinet, 'for the Government to force them further into the red. I have already prevented them earlier this year from making a general price increase designed to reduce the accumulated deficit.' The government's dilemma was clear as 'apart from the statutory duty laid on the Board of making ends meet financially, the obvious course for any enterprise is to increase prices to meet costs provided that its products are still in demand.' The problem was that the NCB was not just 'any enterprise.' Increasing coal prices before the local government elections was not thought politically wise and a 'further rise in the price of coal will be an unpleasant burden on industrial users...as well as on domestic consumers.' On other hand, delay 'will only serve to pile up the deficit which will have to be wiped out sometime.'[119]

A Conservative government pledged to financial discipline in the public sector could not ignore the deficit but, paradoxically, not reducing the deficit might serve to bolster the NCB's position against the NUM. Ministerial involvement impinged on the NCB's statutory obligations and 'if they prevented the Board from taking the action which they considered necessary' ministers would become responsible for managing the industry. In this lay the seeds of a major political crisis as 'certain members of the Coal Board might resign and there might be serious agitation in some coalfields where the miners already felt that the Government were, for political reasons, refraining from charging the steel companies an economic price for high-quality coking coal.' One way of reducing pressure for price increases was to point out to the NUM that 'by continually forcing up the price of coal, they were providing industry and householders with a sharper incentive to adopt substitutes for coal and thereby increasing the prospects of eventual unemployment in the coalfields.'[120]

Ministers were constrained because 'Coal prices were prescribed by the NCB, and such control as the Government exercised rested on an informal arrangement with the NCB.' Moreover, it was not clear whether coal prices were inflationary or deflationary in their wider economic effects as the Chancellor, R.A. Butler, frequently pointed out. Increased coal prices could absorb excessive purchasing power but could also add to inflationary pressure by increasing energy costs. Butler felt that Lloyd's objective of putting the NCB into surplus might create pressure from the NUM for higher wages which would be inflationary.[121] Ministers accepted a 12.5 per cent increase in industrial coal prices but none, other than for the very highest grades, for domestic fuel. The Cabinet consensus was that on balance coal price increases were inflationary even though they 'might secure some economy in the use of coal [but] would in the long run tend to increase the inflationary pressure in the internal economy.' Increases in coal prices would trigger increases in the price of gas, electricity, manufactured goods, and transport all of which would encourage calls for increased wages. The NCB's deficit could not be allowed to accumulate indefinitely, but ministers accepted that resolving this 'might lead the miners' representatives to increase their demands.'[122]

Butler proposed he make it made clear in a Parliamentary statement that coal prices were the responsibility of the NCB not the government's. This suggestion caused serious misgivings in Cabinet as it 'would suggest that the increased coal prices were due to an initiative by the Government, and Ministers would thus assume an unnecessarily large share of the responsibility for the rise in the cost of living'. Nevertheless, coal prices would have to be increased soon as any delay would cost the NCB £22.5m per week.[123] After consultations with the NCB Lloyd reported that the Board were prepared to take full responsibility for the price increases but some ministers remained doubtful about the scale of the increase which seemed to be too large for 'the wider economic purpose of encouraging economy in the consumption of coal and supporting the Chancellor's other efforts to curb inflation.' Lloyd agreed a surplus would increase pressure from the NUM but countered with the argument that large price rises could be justified on purely commercial grounds. The Cabinet agreed in principle with a price rise but could not agree on the amount other than that it 'should not exceed what was strictly necessary to cover increased costs [and] account should be taken on the risk that profits earned by nationalised industries would be dissipated in increased wages.'[124] Lloyd and Houldsworth discussed extensively the size of increase and also how this joint decision could be presented to the media and NUM as an NCB decision. The formulation agreed was that the NCB had 'taken into account' the government's views. The effect on the NUM's wage demands was now a more important consideration than the macro-economic effects of any coal price increase, as Norman Brook minuted Churchill.[125] The Cabinet agreed to ask the NCB to give 'full consideration to all the problems involved' in increasing coal prices. Ministers did not intend to dictate pricing policy or increases 'necessary on commercial grounds' but would inform the Board 'that the total effect...should not be such as to result in a conspicuous surplus which might sharpen the demand for wage increases.'[126]

A decision was delayed by the 1955 General Election. Lloyd's discussions with Hyndley enabled him to avoid having to accept or reject a *formal* request for a price increase from the NCB but a decision could be delayed no longer. Lloyd told his colleagues:

> We could hardly compel the Board to go on increasing their deficit or to postpone indefinitely the discharge of their statutory duty to make ends meet taking one year with another. Moreover, a large increase in the price in the price of coal would incidentally lend powerful support to the Government's fuel policy at two points: it would encourage the substitution of oil for coal and it would encourage the more efficient use of coal.[127]

The bad news was that the Board wanted a 20 per cent increase in coal prices. Butler complained to Eden, the new Prime Minister, that coal was absorbing too much of his and the Economic Policy Committee's time and argued the government had no choice other than to sanction a large increase to enable the NCB to meet its statutory obligations. He pointed out that the UK was now a net importer of coal so a substantial increase would help reduce domestic consumption and reduce import costs and Butler concluded the price rise would have no major consequences for inflation. Brook reiterated these points adding that a substantial rise in coal prices would encourage fuel switching which 'is the only way in which we can reduce the extent to which the national economy is at the mercy of the miners.'[128]

On the following day the Cabinet discussed the proposed increase of 20 per cent in pit-head prices. The principle of an increase had been accepted as had the view that the amount should do no more than cover the NCB's increased costs but even with a 20 per cent increase the accumulated deficit would continue to increase. The Cabinet preferred 18 per cent. A large increase would obviate the need for further increases and 'frequent changes in the price of coal were inconvenient, both politically and industrially' so a large increase would promote stability, improve the Board's finances, encourage greater fuel economy, and promote the shift from coal to oil but an increase of this size was bound to have an inflationary effect. The Cabinet agreed to 18 per cent.[129]

The NCB accepted ministers defined the public interest. J. Latham, NCB Deputy Chairman, admitted to the Select Committee on Nationalised Industries that 'any board would prefer freedom to ties of any kind, but I do not want there to be any misunderstandings about my answer in this sense, that we are a nationalised industry, we have public responsibilities and we do not seek to evade those. From time to time we may think we ought to have more freedom than we get, but we do accept there are public responsibilities.'[130] The NEC accepted the NCB would never openly challenge the government even though it believed 'there was no reason why the Board should be so apprehensive about increasing the price of coal. The policy of the present [Conservative] Government was to increase the price of many commodities by a considerable amount.' Neither would the NEC accept the sanctity of the NCB's balance sheet as coal was subsidising other industries as 'It was fallacious for the Board to argue that "you could not get a pint out of a half pint pot". It lay in the hands of the Board to decide the size of the "pot".'[131] The NCB admitted some coal (coking coal and steam coal for power stations) was sold below production costs but that sold to railways, the gas industry, engineering and house coal was profitable. Furthermore the industry was facing greater competition from foreign coal and especially oil so increased coal prices were 'contrary to long-term Government policy.' The Board freely admitted it was in 'great difficulties' on coal prices but could not justify any change of position.[132] This understandably annoyed the NEC as 'Millions of pounds were being made by other industries largely because they were getting coal at practically the cost of production. If the coal mining industry was to be of a service to the country, selling coal to other industries at prices well below what it could be sold at on the free market, then *the Government must consent to the Board* finding other means of meeting reasonable [our] demands'.[133] The Ministry of Fuel and Power argued 'coal prices are so important that it is a matter in which the Minister must be interested' but decisions were a shared responsibility rather than a ministerial ukase. Ministers judged a request on three grounds: the quality of the case for an increase, the degree to which the proposed increase was self-financing through increased productivity, and its effect on the NCB's deficit.[134]

The NCB were not frustrated free-marketeers. Ashworth wrote:

> There is no doubt that the members of the national board were deeply conscious of their role as providers of a public service...Even though they wanted higher and quicker price increases than they got, they were reluctant price raisers and would never have contemplated charging at world price levels in their first ten years, even if the government had not excluded the possibility.[135]

The Board and ministers shared a common conception of coal's place within the 'national interest.' Sir John Maud stressed coal's function as the foundation of the economy and that the gas and electricity industries 'are very largely tied up with the price of coal, and therefore to leave a monopoly producer in the position of simply deciding for itself what was fair regarding a [price] rise...is of such importance to a large section of the community that the Minister cannot disinterest himself in it'.[136] Ministerial influence over coal prices was also justified as a means of encouraging greater efficiency in the coal industry by preventing the NCB, and by extension, the NUM exploiting their monopoly position by passing wage increases directly to the consumer. For this reason the Select Committee recommended putting the Gentlemen's Agreement into a statutory format, a recommendation rejected by the government.[137]

By the mid-1950s the NUM and NCB relationship had settled down. The NCB retained control over major economic and technical decision making and although the NUM was consulted, its views were not accorded the same weight and status as 'expert' opinion. The NUM was well aware of this and it was disquieted by the failure of the NCB to develop an ethos or managerial cadre significantly different from that of private enterprise. Whilst the NCB was a major innovation the elements of continuity with the private industry within it were as important as the changes, indeed these changes often reinforced existing attitudes. So once allowance has been made for improved material conditions, the reconstruction of the industry, and the existence of consultation and conciliation procedures, the distribution of power in the industry had not significantly shifted and the role of government was to reinforce management's position.

Notes

[1] 5s *H.C. Debs* 423, 20 May 1946, col. 132-33. R. Page Arnot, *The Miners: One Union, One Industry. A History of the National Union Mineworkers 1939-1946*, George Allen and Unwin 1977, pp. 130-60 covers the parliamentary debates in great detail.

[2] Ministry of Fuel and Power, *Coal Mining. Report of the Technical Advisory Committee* CMd.6610, March 1945, para 690-91, p.116. Emphasis added.

[3] *NUM (EC)*, 20 September 1945.

[4] Appendix III. Meeting of the National Joint Committee of the NUM and NACM, 19 December 1945, *NUM (EC)*, 14 January 1946; Appendix I. Production Sub-Committee and Report of Conference of Production Officers, 1 February, *NUM (EC)*, 14 February 1946.

[5] Unofficial Stoppages, *NUM (EC)*, 12 June and 6 July 1947. Under private ownership it had been common to sue unofficial strikes for breach of contract.

[6] W. Lawther's Presidential Address in *NUM (ACR) 1947*, p. 19.

[7] *NUM (ACR) 1947*, p.72. Horner's speech was reproduced as a pamphlet and distributed throughout the industry.

[8] There is no definitive account of Grimethorpe's political and industrial significance. Accounts can be found in NCB, *Annual Reports and Accounts 1947*, paras 74-78, pp. 18-28, N. Fishman, 'The Beginning of the Beginning: The National Union of Mineworkers and Nationalisation', in A. Campbell, N. Fishman and D. Howell eds., *Miners, Unions and Politics, 1910-1947*, Scolar Press 1996, pp. 288-90, and B.J. McCormick, *Industrial Relations in the Coal Industry*, Macmillan 1979, pp. 180-81.

[9] Holmes, who had been an Army boxing champion, offered to fight strikers on condition that if he won the men would return to work. This interesting innovation in industrial relations was not acted upon.

[10] *The Times*, 9 September 1947.

[11] *NCC* (6), 9 September, *JNNC* (22), 26 November, and *NUM (EC)*, 3 December 1947.

[12] Report of the Joint Committee on Production, *NUM (EC)*, 17 November 1948 and *JNNC* (43), 12 January 1949. Edwards had been pressed by Shinwell and the NEC to join the NCB where he was the board member for 'labour'.

[13] *NCC*, 8 March 1949.

[14] *NUM (EC)*, 12 May 1949, and Appendix II Note of An Informal Meeting, *JNNC* (Unions' Side), 20 May 1949. The dispute was over concessionary fuel allowances.

[15] Concessionary Coal, *NUM (EC)*, 22 September 1949. There were also strikes in Scotland protesting at low pay.

[16] *NCC*, 13 September 1949 and *CINC* (23), 14 March 1950.

[17] *CINC* (24), 11 July, *NCC*, 12 September, and Situation in the Industry, *NUM (EC)*, 30 November 1950.
[18] Appendix II. Unofficial Stoppages, *CINC* (31), 22 May 1951.
[19] SI(M) (48) 19, 17 March 1948. Progress Report on Socialised Industries. Memorandum by the Lord President of the Council, para 28, p.5. *CAB 134/689.*
[20] NUM, *Report of the National Executive Committee*, May 1947, p. 217.
[21] *NUM (ACR) 1947*, p. 23 and p. 29.
[22] Resolution 29. Nationalisation and Future Development of the Mining Industry, *NUM (ACR) 1947*, pp. 96-100.
[23] F. Zweig, *The Men In The Pits*, Victor Gollancz 1948, p. 164.
[24] Ministry of Fuel and Power (Coal Division) Circular No D.L.6, 19 October 1942, Conciliation and Arbitration. *LAB 10/895*.
[25] Zweig, *The Men In The Pits*, p. 166.
[26] NUM, *Report of a Special Conference*, 14 March 1947, pp. 3-6.
[27] *NUM (ACR) 1955*, p. 57.
[28] Sir Frank Tribe to Sir Thomas Phillips, 9 July 1942. *LAB 10/895*.
[29] *NUM (EC)*, 14 March 1946.
[30] A.E. Stillwell to all R.I.R.O.s, 15 January 1947. *LAB 10/895*.
[31] *NUM (EC)*, 12 February 1948.
[32] *NCC*, Minutes of the 9th Meeting, 9 March 1948.
[33] Letter from Lord Hyndley to Arthur Horner, 19 April 1948.
[34] *NUM (ACR) 1948*, pp. 52-53.
[35] *NUM (ACR) 1948*, pp. 57-58.
[36] Committee on Consultative Machinery, *NCC*, 13th Meeting, 9 November 1948. The NCC was renamed the Coal Industry National Consultative Committee (CINCC), the conciliation equivalent was the Joint National Negotiating Committee (JNNC).
[37] SI(M) (47), 12th Meeting, 27 November 1947, pp. 4-5. *CAB 134/688.*
[38] N. Dennis, F. Henriques and C. Slaughter, *Coal Is Our Life. An Analysis of a Yorkshire Mining Community*, 2nd. ed., Tavistock 1969, p. 97.
[39] A.J. Taylor, 'The Politics of Labourism in the Yorkshire Coalfield, 1926-1945', in Campbell, Fishman and Howell eds., *Miners, Unions and Politics*, pp. 232-37.
[40] Attitudes to the Nationalisation of Coal, June 1948, p. 28 and p. 29. *Mass Observation File Report 3007.* Hereafter MO FR.3007. MO's survey was conducted in the Doncaster area of the Yorkshire coalfield and looked in depth at two pits (Thorne and Rossington). Questionnaires were distributed to 50 miners and 50 miners' wives, supplemented by reports from MO personnel. The result corresponded to similar findings from South Wales.
[41] Zweig, *The Men In The Pits*, p. 152 and p. 155.
[42] *MO FR.3007*, pp. 21-22.
[43] M. Cole, *Miners and the Board* Fabian Research Group. Research Series 134, Fabian Society Publication/Victor Gollancz, May 1949, pp. 11-12.
[44] *NUM (EC)*, 22 August 1946.
[45] The Miners on the Hearth (May 1948), p. 9. *Mass Observation File Report 2297.* Hereafter, FR 2297.
[46] *FR.3007*, p. 13.
[47] Williams, *A Prime Minister Remembers*, p. 92.
[48] *NUM (ACR) 1949*, p. 146.
[49] J. Bullock, *Them and Us*, Souvenir Press 1972, p. 134.
[50] *Gaitskell Diary*, 21 June 1949, p. 114.

51 FR.3007, p. 27.
52 Cole, *Miners and the Board*, p. 5
53 Resolution 14. Appointments Made by the NCB, *NUM (ACR) 1949*, p. 127.
54 SI(M) (49) 38, 25 June 1949. Suggested Action to Counter Criticisms of Nationalised Industries. Memorandum by the Minister of Fuel and Power, para 7(3), p. 3. *CAB 134/690*.
55 Cole, *Miners and the Board*, pp. 7-9.
56 *NUM (EC)*, 19 December 1946.
57 NUM, *Report of a Special Conference*, 20 December 1946, pp. 25-26.
58 SI(M) (49) 35, 25 June 1949. Suggested Action to Counter Criticisms of Nationalised Industries. Memorandum by the Minister of Fuel and Power, para 7(1), p. 2. *CAB 134/690*.
59 *NUM ACR 1947*, p. 30.
60 *MO FR.3007*, pp. 24-25.
61 NUM, *Report of a Special Conference*, 20 December 1946, pp. 27-30.
62 SI(M) (49), 25 June 1949. Suggested Action to Counter Current Criticisms of Nationalised Industries. Memorandum by the Minister of Fuel and Power, para 7(1), p.2. *CAB 134/689*.
63 NCB, *Annual Reports and Accounts for 1952*, para 10, p. 18.
64 Sir Stafford Cripps quoted in *The Times*, 28 October 1946.
65 Ashworth, *The History of the British Coal Industry*, pp. 122-25.
66 Bullock, *Them and Us*, p. 132.
67 SI(M) (48) 19, 17 March 1948. Progress Report on Socialised Industries. Memorandum by the Lord President of the Council, para 9, p. 2. *CAB 134/698*.
68 F. Williams, *A Prime Minister Remembers*, Heinemann 1961, p. 92 and SI(M) (47) 11th Meeting, 19 November 1947. *CAB 134/688*.
69 J. Gormley, *Battered Cherub*, Hamish Hamilton 1982, p. 38.
70 Bullock, *Them and Us*, p. 134.
71 B.L. Coombes, 'One Year of Nationalisation', *Fortnightly Review*. January 1948, p. 53.
72 Cole, *Miners and the Board*, p. 9.
73 FR.3007, p. 11.
74 *Gaitskell Diary*, 18 July 1948, p. 76.
75 *Gaitskell Diary*, 18 June, p. 70 and 18 July 1948, p. 78. Wage negotiations between Monty and the NUM are one of the great missed spectacles of post-war British politics.
76 *The Times*, 23 November 1948.
77 *Gaitskell Diary*, 23 April, p. 61 and 7 May 1948, p. 68. On 11 May Reid had attended a Board meeting and gave no indication of his intention to resign. Reid's analysis of the industry was very close that of leading Conservative critics such as the ex-coal owner Col. C.G. Lancaster MP.
78 Williams, *A Prime Minister Remembers*, p. 92.
79 A. Horner, *Incorrigible Rebel*, MacGibbon & Kee 1960, pp. 182-83.
80 A. Moffat, *My Life With the Miners*, Lawrence & Wishart 1965, pp. 87-8.
81 *NUM (EC)*, 14 February 1946. Edwards' appointment had been made in March 1946.
82 *NUM (ACR) 1948*, p. 170.
83 NUM, *Report of a Special Conference*, 29 December 1949, p. 1.
84 Coombes, One Year of Nationalisation, p. 47.
85 FR.3007, p. 2.
86 FR.3007, p. 8.

87 NCB, *Annual Reports and Accounts 1949*, para 311-313, pp. 81-2 discusses the impact of unofficial disputes on output. Absenteeism accounted for an average loss of 13 per cent of all shifts in this period.
88 These losses were the product of the tensions inherent in pit-level bargaining and the daily demands of coalmining. Their inspiration was economistic and the high level of conflict in the pits should not be regarded as a 'resistance movement' at the point of production.
89 Ashworth, *The History of the British Coal Industry*, p. 595.
90 L.J. Handy, *Wages Policy in the British Coalmining Industry. A Study of National Wage Bargaining*, Cambridge University Press 1981, p. 215.
91 Handy, *Wages Policy in the British Coalmining Industry*, p. 216.
92 For Yorkshire see B.J. McKormick, 'Strikes in the Yorkshire Coal Industry', in M. Kelly and D.J. Forsyth eds., *Studies in the Coal Industry*, Pergamon Press 1969, pp. 171-98 and Ashworth, *A History of the British Coal Industry*, pp. 509-602.
93 Data derived from various issues of the NCB's *Annual Reports and Accounts*.
94 Zweig, *The Men in the Pits*, p. 70.
95 Dennis, Henriques, and Slaughter, *Coal Is Our Life*, p. 76.
96 *FR 2297*, pp. 6-8.
97 *NUM (EC)*, 29 April and 4 July 1948.
98 National Coal Board Administration, *NUM (ACR) 1948*, pp. 146-52.
99 *NUM (ACR) 1948*, pp. 153-56.
100 *NUM (ACR) 1948*, pp. 157-58.
101 *NUM (EC)*, 7 April 1949. Replies were received from Areas which covered 30% of the NUM's membership.
102 SI(M) (49), 25 June 1949. Suggested Action to Counter Current Criticisms of Nationalised Industries. Memorandum by the Minister of Fuel and Power, para 9, p. 5. *CAB 134/689*.
103 Appendix X. Structure and Conduct of Nationalised Industries. Interim Report Presented by the National Union of Mineworkers, *NUM (EC)*, 23 June 1949.
104 *NUM (EC)*, 16 October 1952.
105 *NUM (ACR) 1949*, pp. 151-69 for the debate.
106 *NUM (ACR) 1950*, pp. 109-11.
107 *CINC*, 27th meeting, 13 December 1951.
108 Minutes of a Special Sub-Committee on the Control of the Mining Industry and National Coal Board, 15 March 1951; and Appendix VI. Minutes of a Special Sub-Committee on the Control of the Mining Industry and National Coal Board, *NUM (EC)*, 10 January 1952.
109 *NUM (ACR) 1952*, p. 40.
110 *NUM (ACR) 1951*, pp. 196-99.
111 Report of the Select Committee on Nationalised Industries (Reports and Accounts), para. 69 (p. xvi). *HC.187-I*.
112 H. Hooper to Eden 2 May 1956. *PREM 11/1746*.
113 Ashworth, *The History of the British Coal Industry*, p. 635. See also the *Report of the Advisory Committee on Organisation*, National Coal Board 1955, para 27, p. 7.
114 SI(M) (48) 39. Current Problems in Socialised Industries. Note of a Meeting at the Bank of England, 3 May 1948, p. 2. *CAB 134/698*.
115 Increased Price of Coal. EA (51) 4th Meeting Minute 2, 19 December, and Coal Prices (51) 19th Conclusions, Minute 1, 20 December 1951. *PREM 11/1746*.

116 *HC 178-I.* Report of the Select Committee on Nationalised Industries (Reports and Accounts), Appendix 4 Coal Prices, pp. 136-37.
117 NUM, *Report of a Special Conference*, 7 December 1951, p. 4.
118 *NUM (ACR) 1953*, p. 71.
119 Coal Prices. C (54) 116 29 March 1954. Memorandum by the Minister of Fuel and Power. Norman Brook to Churchill, 30 March 1954. *PREM 11/1746.* Lloyd was appointed in October 1951 and served as minister until December 1955. He had considerable familiarity with energy policy; he was Minister of Mines (April 1939 to May 1940), petroleum minister (May 1940 to June 1942), and Parliamentary Secretary at the Ministry of Fuel and Power until May 1945.
120 CC (54) 23rd Meeting Conclusions Minute 6, 31 March 1954. *CAB 128/27.*
121 EA (55) 7th Meeting. Minute 2, 23 February 1955. *PREM 11/1746.*
122 CC (55) 17th Meeting Conclusions Minute 1, 23 February 1955. *CAB 128/28.*
123 CC (55) 18th Meeting Conclusions Minute 1, 24 February 1955. *CAB 128/28.*
124 CC (55) 19th Meeting Minute 11, 2 March 1955. The discussion continued at the 7 March Cabinet. *CAB 128/28.*
125 Coal Prices C (55) 61, 7 March 1955. Memorandum by the Minister of Fuel and Power. Norman Brook to Churchill 7 March 1955. *PREM 11/1746.*
126 CC (55) 22nd Meeting Conclusions Minute 5, 9 March 1955. *CAB 128/28.*
127 CP (55) 45, 22 June 1955. Coal Prices. Memorandum by the Minister of Fuel and Power, para 4. *CAB 129/28.*
128 Butler to Eden, 22 June, and Brook to Eden. 22 June 1955. *PREM 11/1746.*
129 CC (55) 17th Conclusions Minute 4, 23 June 1955. *CAB 128/28.*
130 *HC.187-I*, Minutes of Evidence, q.90, p. 18.
131 Appendix III. Special Meeting of the NUM and the NCB, 12 November 1953, *NUM (EC)*, 12 November 1953.
132 *NUM (EC)*, 6 January 1954.
133 Wages-Meeting with the NCB, *NUM (EC)*, 7 January 1954. My emphasis.
134 *HC.187-I*, Minutes of Evidence, q.243 and q.244, p.40. Evidence of Sir John Maud and R.J. Ayres (Ministry of Fuel and Power).
135 Ashworth, *The History of the British Coal Industry*, p. 579.
136 *HC.187-I*, Minutes of Evidence, q.269, p. 43.
137 *HC.187-I*, Report of the Select Committee on Nationalised Industries (Reports and Accounts), paras 88-89, p. vxiii.

Chapter 4

Conservatives and the NUM

Introduction

The election of a Conservative government would inevitably cause tensions in the NUM – government relationship. Not only were Conservatives hostile to nationalisation but the legacy of 1926 and the responsibility of Conservative dominated governments for the woes of the inter-war years exerted a strong influence in NUM politics. Churchill was personally reviled in many mining communities as the Liberal Home Secretary who deployed troops against striking miners at Tonypandy in 1909 and subsequently as Chancellor of the Exchequer in Baldwin's government for being the leading Cabinet 'hawk' in 1926.[1]

Many Conservatives had little sympathy for the miners. During the unofficial strikes of early 1944 one backbench MP had written; 'would to God that they could be treated by Russian methods – but what the Russians may do if done here would arouse a tremendous uproar, from the Archbishop of Canterbury to Mr Shinwell would come shrieks of dismay, and yet the shooting of a few miners would end strikes.'[2] Significantly, the Conservative attack on mines nationalisation was circumspect and 'in 1945 the party had opposed nationalisation in principle, while making exceptions in practice; in 1951 it opposed nationalisation in practice but claimed to be pragmatic in principle. Neither position entailed radical action.'[3] The 1945 Conservative manifesto argued for a new start in the industry based on the Reid Report which would remain privately owned.

On several occasions Arthur Horner suggested that faced by a Conservative government the NUM 'would perhaps have to use every weapon in our armoury.'[4] Such statements were described by Churchill as 'part of the Communist conspiracy to bring Britain under the heel of the Kremlin.' He conceded that the mineworkers had a right to the best possible wages and conditions, 'But now the Communist Horner has stepped outside the sphere of industrial disputes and threatens the whole British democracy, 30 million voters, with a national strike to bring the country down if they dare express their opinion and wishes at the polls.

This is an insult to the will of the people which no democracy could endure.'[5] This shadow boxing was part of the political game and the NUM and the Conservative leadership had domestic audiences to keep happy but this rhetoric was for neither side a determinant of policy or relations.

The Conservatives and the Coal Industry

Conservatives did not, at least in public, blame disruption and unrest in the pits on the mineworkers but on state control and the frustrations caused by NCB bureaucracy. 'The transfer of the ownership of the industry', the Conservative Party argued, 'from millions of private investors to the State accomplishes nothing except an increase in the totalitarian power and patronage of Socialist ministers'.[6] The 1947 Grimethorpe Dispute was emblematic of the new order and proved that:

> nationalisation offers no solution for the troubles affecting the coal industry. It is very doubtful if such a state of affairs could have arisen when the mines were privately owned because the colliery owners as businessman were prepared to take advice from their managers on the spot. The Coal Board, on the other hand, is already a branch of the Civil Service, a vast bureaucratic organisation which is called upon to make large-scale decisions at central headquarters far removed from coal-face conditions.[7]

The Conservatives' problem was that this rhetoric could be easily portrayed as an attack on public ownership and because of the country's total dependence on coal and the NUM Conservatives could not afford to be seen to be even hinting at confrontation with the miners. Churchill told the 1947 Party Conference that:

> We are quite sure that the nationalisation of our industries will not make them profitable to the country or satisfactory for their workers. It is not in the interests of the ordinary wage-earner to be the servant of the State, that is, to be the servant of an all-powerful employer centralising the management of nationalised industries in the public departments of Whitehall.[8]

Even so he would not commit a future Conservative government to returning the coal industry to the private sector. R.A. Butler declared that 'I think it is very natural that the Party should desire to have complete liberty to restore to private ownership industries which are under the threat of nationalisation, the vesting date for which has not been placed.'[9] As coal's vesting date had passed Butler's statement meant coal would not be denationalised.

Churchill restated frequently the Conservative belief that coal nationalisation had failed the country; 'The Coal Board boasted the other day that they had made a nominal profit of a million in their last year's working of the coal mine...Anyone can make a profit if they have a monopoly'.[10] Conservative critics were interested in decentralisation to encourage competition between sub-units. *The Industrial Charter* (May 1947) reiterated the party's principled opposition to nationalisation but this opposition would be expressed pragmatically; 'Rather than proposing the complete denationalisation of the coal industry, the methods by which the Socialists have tried to run it should be examined and modified.'[11] This became party policy in 1949.

In a radio broadcast (23 July 1949) Anthony Eden pledged that 'We will undertake no further nationalisation. We will restore free enterprise where this is practicable'. This commitment applied specifically to iron and steel and road transport but 'Where we have no alternative but to leave some industries nationalised, we shall radically overhaul their administration and seek to make them more efficient, less centralized and more human.'[12] At the 1949 Conservative Conference Churchill applied Eden's statement explicitly to the NCB by accepting that 'It is physically impossible to undo much that has been done. You cannot thrust the coal mines and the railways back upon the private owners. They would not take them. All that can be done in these two basic services is to decentralize and cut down on the enormously swollen costs of management.'[13] *This Is The Road*, the 1950 Conservative manifesto, promised that 'We shall drastically reorganise the Coal Industry as a public undertaking by restructuring the National Board and by giving autonomy to the Area Boards. By decentralising the work of the National Board we shall give greater responsibility to the men on the spot and revive local loyalties and enthusiasm.' This was opposed implacably by the NUM as backdoor denationalisation and in a parallel to Labour's lack of detailed preparation for nationalisation, the Conservatives did not detail the form decentralisation might take. For the 1951 election the party adhered to decentralisation but with a significant change of emphasis; '*Coal will remain nationalised. There will be more decentralisation and stimulation of local initiative and loyalties, but wage negotiations will remain on a national basis*'.

In November 1951 Charles Gray, a Labour MP, asked Geoffrey Lloyd, the Minister of Fuel and Power, about government policy on the NCB's structure. Lloyd replied that primary responsibility for its structure lay with the NCB and should ministers think any changes were necessary they 'would only be taken after full consideration with those concerned' including the NUM. He then 'made it clear' neither the Government nor NCB intended to revert to district wages agreements or inter coalfield competition. Gray referred to the miners' 'deep suspicion' of the Government and called on Lloyd to dissociate the Conservative government from all references to decentralisation in the Conservative manifesto. Lloyd refused as his statement was in full accord with the manifesto and Gaitskell's statement of November 1948 expressing support for the greatest degree of decentralisation practicable.[14] Suspicion of the government's intentions surfaced periodically. In March 1953 a Labour backbencher, Stephen Swingler, asked Lloyd what directions on financial and managerial decentralisation the government had given the NCB since 1951 and what directions the government intended to give. In response to Lloyd's answer ('None'), Swingler asked 'if he will make it clear to some of his political friends that he is not in favour of a reversion to district wage agreements or the setting of one district against another in competition; and that the propaganda which some of his friends are carrying out in the coalfields is very disruptive to the present operations of the Coal Board and the [NUM]?' Lloyd repeated that the definitive government position remained that of November 1951.[15] Nevertheless, some Conservative backbenchers remained incredulous that Conservative ministers were running state industries and 'The extraordinary thing is that we on this side of the House are defending a policy in which we never believed and which we never thought could succeed'.[16]

A common view is that after its defeat in 1945 the 'overwhelmingly benevolent if paternalistic post-war attitude towards industrial relations reflected the Conservative leadership's recognition that the new power and authority of the trade unions acquired during the 1940s could neither be denied or reversed.'[17] In the 1953 Debate on the Address Churchill commented 'It may sometimes be necessary for Government to undo each other's work, but this should be an exception not the rule'. Consequently, although Conservatives opposed public ownership, where it was retained, as with the mines, 'we have done and are doing our utmost to make a success of it.'[18] This is portrayed as the appeasement of the unions to win the unions' cooperation and particular opprobrium is heaped on Walter Monckton, Churchill's Minister of Labour, for failing to deal with the 'union problem' despite receiving explicit instructions from Churchill not to antagonise the unions.[19]

Monckton recalled 'Winston's riding orders to me were that the Labour Party had foretold grave Industrial trouble if the Conservatives were elected, and he looked to me to do my best to preserve Industrial peace'.[20] An attempt to deploy 2,400 Italians in the mines foundered on 'firm and growing' opposition from the NUM at local level and Lloyd 'reluctantly reached the conclusion that we had much more to lose than to gain by bringing further pressure to bear on British miners in this matter.'[21] Civil servants argued the scope for government action in a major dispute involving the coal industry was limited; 'Not only might absolutely essential services (such as hospitals) have to be suspended, but the hardship, especially to the very old and the very young, of living without warmth and light at the coldest time of the year might be intolerable. This, indeed, may be the limiting factor in deciding whether in any practical sense strikes in these key sectors could be "borne" by the community.'[22] Monckton was pressed by some ministers and the Conservative Party to ban the closed shop but this was a sensitive issue in the mines. Harold Macmillan related to Monckton a conversation with Sam Watson at a dinner given by Peterlee New Town Corporation. Watson asked Macmillan to 'Tell your friend Sir Walter Monckton, not to give in an inch about the Closed Shop issue. If you do, we are all sunk.'[23] Legislation on the closed shop would have provoked a furious reaction which moderate-constitutionalist leaders (such as Watson) would be unable to restrain and this unrest might engulf both the unions and the Conservative government.

The problem was that 'The Conservative Party came into power at a time of crisis when the economic position was such that it was considered essential that no large industrial dispute should take place.'[24] The change of government did not transform the fuel supply situation. In November 1951 Lloyd warned the Cabinet that this was 'exceedingly grave' as the country needed 19m tons of coal to get through the winter but only 17m tons were in stock. Lord Leathers, the Secretary of State for Co-ordination of Transport, Fuel and Power, described a 'difficult legacy' and 'a certainty of a breakdown in supplies to the public' unless 500,000 tons of coal was imported from the United States. Churchill was most displeased and instructed ministers to make it clear to the press and party that this was the result of Labour's lack of foresight but to emphasise that the government attached no blame to the mineworkers.[25]

Vulnerability encouraged ministers to seek alternative fuels. Lloyd pointed excitedly to the possibilities of the underground gasification of coal. This, Lloyd enthused, would permit the exploitation of large quantities of deep poor quality coal *'above all*, without the need of miners.'[26] Nuclear power was regarded by ministers as the long-term solution to Britain's dependence on coal. In 1952 Lord Cherwell minuted Churchill, 'You will be glad to learn that there is at last a prospect of obtaining power from nuclear reactors.' Cherwell believed 'the whole of the 35 million tons of coal now used yearly...for the generation of electricity could be replaced by considerably less than 100 tons of uranium', but achieving this would take many years at great financial cost. By the end of 1954 ministers were close to a final decision on initiating a civil nuclear power programme which was not expected to cause trouble with the miners as it would supplement, not replace, coal.[27] The most significant long term policy decision was to promote the substitution of oil for coal. Lloyd was 'profoundly disturbed by the continuance of the chronic shortage of coal which gives the miners such a stranglehold on our economy.' Any increase in output from the mines would be absorbed by economic growth so government faced 'a prospect of continuing shortage, no increase of exports, no derationing of house coal and our remaining indefinitely at the mercy of the miners.' Lloyd's solution was to 'ease the tension' at the margin by deploying the 'expansive power of the oil industry' to supply 3m-4m tons of fuel oil to displace 8m tons of coal, increasing this to 10m tons (20m tons of coal).[28] Until the flood of cheap oil in 1957 the basic political and economic fact was that the Conservatives depended on coal and the miners, however the industry's failure to supply the energy required and the country's vulnerability to the disruption of oil supplies as demonstrated by the 1956 Suez Crisis, led the government to encourage the expansion of oil refining and improve the distribution system.[29]

The Conservatives returned to office in 1951 to face a major economic crisis which it met by a policy of deflation. For the unions increases in money wages had become the measure of success but few state industries (including coal) were generating a surplus sufficient to satisfy their employees' expectations, unlike the booming private sector. This produced 'a dynamic process capable of quite different economic effects from the mild wage-inflation of the late 1940s.'[30] Politics were suffused by a fear, exacerbated by the Cold War and anti-communism and the resilience of Labour support at the polls, that incoming Conservatives would face a wave of politically motivated strikes. Conservatives acknowledged that the TUC and individual union leaderships (including the NUM) opposed government policies but argued political strikes would break the union movement. The Conservative Research Department concluded that despite significant Communist influence in several coalfields and isolated outbreaks of industrial protest over government policy (in, for example, the South Wales coalfield over the 1952 budget) the unions, individually and collectively, would work with the government. CRD concluded 'It would be impossible to maintain there is any widespread industrial action for political purposes.'[31] The TUC General Council placed its weight behind accepted constitutional practice reiterating that 'It is our long standing practice to work amicably with whatever government is in power and through consultation jointly with Ministers and the other side of industry to find practical solutions to the social and economic problems facing the country. There need be no doubt therefore of the attitude of the TUC towards the new government.'[32] As a TUC affiliate the NUM was bound by this as well as by the demands of public ownership.

Churchill and Monckton enjoyed good personal relations with the older generation of union leaders (including Lawther) and recognised that confrontation would damage the Conservatives at a time when a resurgence of class conflict seemed a distinct possibility and when the Conservatives were still blamed widely for the mass unemployment 1930s.[33] One of his private secretaries believed Churchill was 'daunted by the spectre of a general strike and of the iron determination and huge organising energy needed to counter it if it came. *Nineteen-twenty six still haunted him*'.[34] This fear of confrontation also permeated the NUM's leadership. 1926 and its aftermath were formative experiences and should never be repeated. The mineworkers' industrial and political experience convinced both the MFGB and NUM that more could be gained at a lower cost from co-operating with the state than by confronting it. 'Whatever we may think', Lawther told the 1952 NUM Conference, 'and however we may feel about the trickery and deceit that was adopted to win power for the Tories in October 1951, it must not be made either the excuse for the alibi for similar tactics.'[35]

The NUM's policy was that after nationalisation strikes were illegitimate and unnecessary but how did the election of a Conservative government change the situation? After 1951 unrest and disaffection in the coal industry were so severe that 'nationalisation could be destroyed and the Industry could be so affected that it might never recover' and a Conservative government might be tempted to use such a situation to attempt to restructure or even, as it was pledged to do with steel, return coal to the private sector. As the NUM 'had advocated the necessity for nationalisation [it] must, therefore, accept responsibility for its success. It was, however, not the sole responsibility of the Union; it was the joint duty of the Board and the Union and its success could only be ensured by cooperation and mutual trust' even though management – union relations had deteriorated.[36] A united front between the NUM and NCB based on their common interest of making a success of public ownership was a major defence against any threat from the new government. This, in turn, required the union restrain its members' proclivity for industrial action and secure their co-operation in management's strategy for reconstructing the industry.

Although the Conservative Party was identified as the party of free enterprise it had never implacably opposed public ownership. Whatever the party's preferences the governmental elite accepted Labour's programme (other than steel and road haulage) and confined itself to statements about the need to make the remainder more commercial and efficient, objectives which had been discussed at length by the Socialisation of Industries Committee. Dependence on coal and the NUM drastically reduced the Churchill government's room for manoeuvre and there were no significant departures from the structures created by the Labour government. One was Churchill's creation of Overlords to co-ordinate related policy areas. On 30 October 1951 Lord Leathers was created Secretary of State of Transport, Fuel and Power. A major problem of co-ordination was that Leathers rejected the idea of a co-ordinated fuel policy in favour of consumer choice, he refused to plan the fuel and power industries and confined himself to facilitating (rather ineffectively) co-ordination via informal meetings between himself and the board chairmen.[37] Leathers continued the pattern inherited from Labour, he had no effect and his post (and those of the other Overlords) was abolished on 3 September 1953. A second innovation was the creation in 1953 of the Select Committee on Nationalised Industries (SCNI) in response to backbench pressure for some say in the affairs of the state industries. Hitherto the main scrutiny mechanisms were the debates held on the annual reports and questions to ministers, but the convention that ministers did not answer questions about day-to-day operations was enforced rigidly. The SCNI investigated the NCB in 1957 and reported in 1958 but did not prove to be onerous.[38]

What infuriated Conservative backbenchers was that by the mid-1950s coal had been publicly owned for nine years, prices had increased substantially and wages improved but coal was still in short supply. This 'expensive flop' could not be denationalised, the long-term solution to the nation's dependence on coal and the mineworkers was using oil and nuclear power to break coal's monopoly but, in the short term, a number of MPs used criticism of the NCB in an attempt to improve its performance.[39] Sir Gerald Nabarro and Col. Claude Lancaster (an ex coal-owner) led a guerrilla campaign against the NCB *via* the Conservative backbench Fuel and Power sub-committee of the 1922 Committee, chaired by Sir Victor Raikes with Nabarro as joint secretary.

Tensions between Conservative MPs, the Coal Board and the NUM came to a head in the spring of 1953. The Cabinet were warned that a Conservative backbencher, Robson-Browne, had tabled a written question asking for an increase in coal prices to be delayed pending an inquiry into the need for such increases. Several other Conservative MPs had tabled a motion deprecating the price increases advocating an inquiry into the structure and administration of the NCB. The Cabinet had already authorised the price increase which could not be withdrawn without damaging Cabinet authority but, more seriously, 'an enquiry would not be well received by the miners and, if it met with any encouragement, was likely to hamper the joint efforts which were now being made by the Coal Board and the [NUM] to secure increased production.' The Cabinet agreed these efforts by a minority of backbenchers to undermine the coal industry should be resisted.[40] Responding to similar complaints, Monckton played a straight bat, praising the joint NUM/NCB efforts to raise productivity but the questioning of the NCB's success immediately raised the political temperature. Shinwell warned Monckton about attempts 'in certain quarters, political and otherwise, to create prejudice against the miners in order to create prejudice against nationalisation.' Monckton dissociated himself and the government from these efforts; George Jeger MP suggested Monckton arrange for Conservative MPs 'to go down a coal mine and do a few shifts'.[41]

Eventually 120 Conservative MPs signed a motion critical of the NCB, called for an inquiry and disputed the government's policy of making nationalisation work. This threatened a huge row in the government and with the NUM, and the situation worsened despite meetings between Lloyd and Leathers and the Fuel and Power Committee. Despite its frequent criticisms of the NCB, any threat, real or imagined, to public ownership caused the NUM to round on the industry's critics. The Conservative backbenchers' call for an inquiry was condemned by the NEC as a 'subtle attempt of the Tory Party to destroy nationalisation' by creating bitterness and division between management and men. The NEC warned that 'Any attempt by the Tory Government to interfere with the national structure of the Industry by decentralisation on a district basis, with all the pre-war disruption of our coal economy will be resisted by the [NUM] with every legitimate means at its disposal.'[42] At the 1953 Conference Tom Stephenson asked the pertinent question, what was meant by resisting threats to the industry by any 'legitimate means'? Stephenson suggested 'legitimate means' confined the NUM to Parliamentary means 'and if that is all we can use, then take it from me, as it is a question of a majority walking into the lobby, then we are sure to be defeated.'[43] With the help of the Whips Lloyd managed to beat off the backbench opposition at a meeting of the 1922 Committee on 23 April and 'the confrontation was widely hailed as a smashing victory for Geoffrey Lloyd.'[44]

Whilst direct Conservative criticisms of the miners and the NUM were comparatively rare, this restraint did not apply to criticising the Board. Even an MP making his maiden speech felt able to return to the theme that the NCB was an impersonal monopolistic bureaucracy responsible for poor industrial relations. Aubrey Jones, on the left of the party and who within a year was to be appointed Minister of Fuel and Power, expressed 'the opinion that the fortunes of the Coal Board are far gone. Whether they can be redeemed or retrieved at all is a matter of the gravest doubt.' Lancaster condemned the NCB's inability to bring new capacity into production and that the industry was far too centralised. 'It is', Lancaster insisted, 'silly to suggest that the industry is not sick and sorry...I still hold very strongly to the view that the industry is not administered and organised to the best advantage.' Knight of the Shires bemoaned the failure of the industry to respond to the nation's need for coal, all the miners needed was good leadership. 'If we are patient,' argued Sir Albert Braithwaite, 'if we get good team work, and if we can attract more and better leaders into the industry we shall solve our problems. The miner is one of the finest citizens we have and will respond magnificently if given the opportunity.' Viscount Lambton, a scion of a Durham coal-owning dynasty, confessed 'this is rather a melancholy occasion. About eight and a half years have elapsed since the nationalisation of the mines and we have come to this debate today rather, as it were, to a wake of the coal industry.'[45]

So grave was the unrest thought to be that Patrick Buchan-Hepburn, the Chief Whip, telegraphed the Prime Minister, Anthony Eden in Geneva; 'Coal Debate. No harm done and good speeches from front bench. No abstentions in division.'[46] The speeches and the Chief Whip's actions testified to the continuing ability of the coal industry to cause tension within Conservative ranks. A significant group of MPs believed that it was a mistake for the state to be intimately involved with such a strategically vital industry, but one which appeared incapable of meetings its obligations and which was dominated by a union with a sensitive relationship with the Conservative Party and governments. On the other hand, the industry was publicly owned and Conservative government had no option other than to run the industry. Sir Gerald Nabarro, one of the government's severest critics, conceded the Ministry of Fuel and Power was 'a very hot seat indeed. It called for a political prodigy: a Tory who could make Socialist state ownership work.'[47]

Wage Bargaining, Industrial Action and Union Strategy

Like its predecessor the Conservative government was deeply concerned about the supply of coal and the effect shortages would have on the economy and the government's political fortunes. Equally, it was concerned about the related issue of mineworkers' pay because the NUM argued that the only way output could be increased was by paying sufficiently high wages to both retain and recruit mineworkers. This was, however, perceived by ministers as having dangerous inflationary tendencies.

The effect of the election of a Conservative government on the NUM's political strategy and orientation concerned many in the union leadership. The left argued the NUM had been mistaken when it determined its strategy on party-political grounds, instead the NUM should pursue its members interests even at the risk of clash with government. Many in the NUM's leadership were appalled by the idea of confronting government, seeing this as a negation of public ownership and offering the Conservatives an excuse to break up the NCB. Horner wrote later:

> We had to be very careful that we did not provide the Tories with an excuse to replace nationalisation by some decentralized form of organisation, and we had to realize that if there were a national strike against the [NCB] that would provide just the excuse that the Tories wanted. And so the policy of the NUM was always to reach and agreement over the table rather than to succumb to pressure for strike action even in cases where the grievances were very great.[48]

In this context, it is important to remember the Conservative sense of vulnerability in late 1951 and early 1952 and that the mineworkers were a major concern of Churchill's.

Wage bargaining was especially sensitive because of the industry's size and because other unions would use miners' wages as a comparator. At the time of the 1951 election the NUM lodged an 18 per cent wage claim and Leathers was anxious the NUM and NCB negotiate a settlement as 'arbitration might be equally costly and would prejudice the subsequent discussions on methods of increasing output.' Raising coal output was the government's main objective but the NUM would not discuss output until pay was out of the way; the NCB had offered 8 per cent but the Cabinet were advised that the NUM would settle for 10 per cent and it agreed to permit the NCB to negotiate on this basis.[49] Lord Cherwell, Churchill's confidante, urged him 'to make the miners a privileged class' and urged the government promote a 'New Deal' for the miners. Cherwell believed that after two weeks of a coal strike the government would have to surrender and although Churchill disagreed with Cherwell's pessimism he nevertheless urged Leathers to concede more pay and privileges to the miners to ensure coal output.[50]

The strategy of the NUM leadership was to insist the NCB as an employer was different from the coal-owners, that the miners had made real gains since 1947, and that the NCB should be defended against any threat. The NUM regarded the NCB as a key element in, and buttress of, the post-war political settlement and believed that its duty was to preserve the NCB by pressure group and parliamentary action rather than industrial action. Lawther urged delegates to:

> Remember that era of wickedness, chaos and strife you and your fathers had in an unplanned industry in 1926? You can never forget it. Over the broken homes and derelict areas the mad men ruled. Never again can we, nor will we, go back to that reign of terror; that iron curtain is broken...

Nevertheless the arrival of a Conservative government was seen by some a political watershed. Jack Watson (Durham) had 'no illusions as to what Tory control would mean...Can anyone here imagine Mr Churchill being given full power or control over our industry and continuing to produce the same improved conditions?'[51] Discontent over pay in the industry encouraged calls for a change in NUM strategy especially as the perception gained ground that the NCB was essentially and increasingly a 'Tory body' enmeshing the NUM in an industrial relations system designed to neutralise the miners' power and 'sooner or later we will have to get out of the tangled web into which we are being spun, and we have to remember the real weapon we possess, and that is our Union's strength.'[52] At the 1951 Conference it was proposed the NUM amend Clause 9 of the Conciliation Agreement and end compulsory arbitration to strengthen the NUM's hand in negotiations. Amending Clause 9 risked 'going back to the old days of struggle and strife.'[53]

Even before the return of the Conservatives the NCB was seeking to operate on conventional business lines. Much of the industry was loss making and the NCB depended on profitable coalfields (notably Yorkshire and Nottinghamshire) subsidising the loss makers. This cross-subsidisation had implications for the NUM's unity and solidarity as 'the time is coming when there has got to be a little *quid pro quo*, and we have to be very careful, because if this Union falls into disunity, we are all sunk, profitable and unprofitable districts'. If the NUM fragmented there would be major political consequences for the industry as 'there are plenty of wolves wanting decentralisation – plenty of them wanting to go back to that. I tell you this: Do not many of you be so proud; there will be only three districts this year that will pay a profit – three out of the lot.'[54] The deterioration in the NCB's finances led Sir Hubert Houldsworth to warn the NEC that 'If private enterprise ran the Mining Industry many of the pits would have to be closed down, for during the first months of 1952 416 collieries producing 98m tons of coal were profitable, whereas 470 collieries producing 59m tons of coal were unprofitable.'[55] Lord Hyndley warned that 'Until the terms of compensation to the former owners have been completed it is *impossible to calculate* the financial commitments of the Board'. Equally worrying was the uncertainty caused by the economic crisis and the impact of the government's deflationary fiscal and monetary policy. The NCB 'cannot isolate itself from the existing financial and economic structure of the country. The Board will be governed by the *normal accepted practice* in its method of accountancy'.[56]

The political complexity of the bargaining climate was demonstrated at a Special Conference called to discuss the continuation of Saturday working where a second resolution was added condemning Butler's 1952 budget. The rationale (extra output) for Saturday working remained, but the agreement had been negotiated with the intention of aiding the Labour government. Now there was a Conservative government, so should the NUM aid a government attacking working class living standards? 'There is a feeling', Horner noted, 'with which we [the NEC] have great sympathy, that reprisals ought to be used against the increases in the cost of living' but ending Saturday working 'does not represent the total armoury of the Union in the struggle against this Government or against the cuts'. The Executive expressed a willingness to fight the Conservative government but not on this issue or at this time. The situation, it argued, called for class-wide action, for if the miners acted unilaterally unemployment would rise and the NUM would be blamed for fragmenting working class unity.[57] Bert Wynn (Derbyshire) questioned this political assessment. The Extended Working Hours Agreement was agreed by the mineworkers 'to protect the Labour Government in order for them to be able to carry out their policy of full employment...I say that particular time has gone' because the Conservatives were attacking full employment.[58] The debate's direction worried Lawther who hurried the discussion to a close; 93 delegates voted for the continuation of Saturday working, 51 were against. On a card vote 413,000 voted in favour, 148,000 opposed but 142,000 abstained.

The second issue discussed related to how the NUM should oppose the Conservative government. The use of industrial action for political purposes was the antithesis not only of the NUM's post-1926 political development but also of public ownership, however as Butler's 1952 Budget was political 'we have got to make our protest on a political issue in a political fashion.' Likening the NUM's power to that of Samson, W.E. Jones insisted that 'We do not want, because the arch priest in the temple is a reactionary, to pull down the temple. What we want to do is to change the arch priest in the temple, and make the temple serve the best good of the people who make up the community of that temple'. Pressure would be exerted via the miners' MPs and by a huge anti-government vote in the May 1952 local elections, the NUM would also seek compensation for the budget's prices rises but it would not sanction strikes against government policy. The resolution was approved.[59]

1952 'was exactly the time when plant bargaining, leap-frogging wage claims, unofficial strikes, and "indiscipline" first began to affect the unions' once-favourable presentation in the press.'[60] The government's information minister, Lord Swinton, instructed all government departments to estimate how many jobs would be lost as a result of inflationary trade union wage demands and then feed the results to the BBC, labour and industrial correspondents. The miners were a key target of this effort.[61] There had been some concern in government that the mineworkers might become part of a general industrial unrest, but ministers quickly appreciated that they could rely on the NUM's constitutionalism. 'We might as well face now', Lawther continued, 'as later this choice of whether, because we lose a General Election, we should then use our industrial power for political objectives.' To do so, he continued, 'would be to say bluntly to the electorate, if we do not have our representatives elected to power, then get ready for another General Strike...a policy that can only lead to national suicide'.[62]

Government made no attempt to control directly wages in the coal industry but the NCB's losses and its request for increased prices offered ministers an opportunity to influence wages indirectly. NCB losses (£8m in 1952 and a projected £36m in 1953) made an increase in coal prices inevitable and in January 1953 Leathers recommended an increase of 10 per cent to the Cabinet. Some Cabinet members wanted the price increase to come as soon as possible after the settlement of the NUM's pay claim so the public would see 'it is they who have to pay for higher wages' whereas others thought this unwise as it risked antagonising the NUM.[63] Leathers believed that the NUM 'had now been brought to realise the grave economic consequences which had been caused by their successive wage demands, short-time working and stoppages. They were now ready to co-operate with the Coal Board in intensive joint efforts to increase productivity.' Other Cabinet members led by Monckton resisted a price rise and urged the deficit be reduced by increases in productivity even if this required short-term subsidies. Subsidies were, however, unpopular with civil servants and other ministers, and would provoke apoplexy amongst the NCB's backbench critics. A substantial increase in coal prices would be resisted by industrial and domestic consumers, and would accelerate the wage-price spiral. Faced by a divided Cabinet Churchill suggested the alternatives be put to 'the responsible leaders' of the NUM. This was not done.[64]

Monckton wanted to create a NCB – NUM, government consensus by trading a 5 per cent increase in output in return for a £9-£10m subsidy. A price increase would affect the cost of living and stimulate compensatory wage demands throughout industry so the industry would have three months to boost output if it failed or the NUM refused 'then the Government in increasing coal prices would have the whole country on their side and would have placed the responsibility far more squarely on the shoulders of the miners'.[65] Cherwell minuted his disquiet to the Prime Minister. He was deeply concerned about any 'suggest[ion] it was their [the miners'] rapacious wage demands' which led to increased coal prices, to do so 'would undoubtedly create just the bad atmosphere [Monckton] rightly fears'. So, 'If no attempt is made to blame the miners' greed and slackness for the increase I cannot see why raising the price should cause bad feelings' and the only alternative (subsidy) was unacceptable. This would be a 'disastrous innovation when we are pledged to reduce Government expenditure. Not only would it go far to nullify all the painful economies we have been trying to make in other fields; it would remove all incentives to keep costs down in this vital industry. Once a Conservative Government sets the example of letting one nationalised industry quarter itself on the Exchequer there will be no end of similar demands which in the end will destroy the national economy'.[66] Conservative backbenchers got wind of the proposed price increase and began to make trouble.

The Cabinet 'recognised the risk that the proposed increase in coal prices would disturb the stability of the national economy and cause an inflationary spiral'. The coal industry had to be dissuaded from believing that price increases would be passed on to consumers and that increased wages would be paid for by increased coal prices. Without a price rise it would need an unheard of 15 per cent annual increase in productivity to eradicate the NCB's deficit but ministers believed the NUM was willing to co-operate in increasing output and there was a strong economic case for increasing coal prices.[67] The NUM and NCB joint campaign to increase productivity as part of the annual wage settlement was seen by the Cabinet as a positive development. Monckton remained gloomy, reminding the Cabinet of the TUC's Economic Committee's warning against any action which might increase the cost of living and he also pointed out that unemployment was rising. Monckton 'could not view the prospects of industrial relations in the latter part of 1953 without considerable apprehension.' Ministers sanctioned a 10 per cent increase in coal prices coupled with a clear warning to the NCB that this was linked to the elimination of the accumulated deficit.[68]

Some ministers believed the government had been too lax. Lord Woolton, the Lord President of the Council, told Churchill 'the time had come for increases to stop and that the Government ought to announce the fact that there were going to be no more increases in prices, and no more increases in wages'. This did not find favour with Churchill.[69] Any government action which made the NCB's financial situation worse might jeopardise the NUM's acquiescence. The NUM's attitude to the NCB's balance sheet was contradictory because while everyone tacitly accepted that the balance sheet was a factor in negotiations, the NUM would not concede this publicly. 'Once', Horner argued, 'we start to argue with the Coal Board, *especially in the life of the Tory Government*, as if we are going to be affected by their balance sheet, we become limited by their balance sheet and that is what we do not intend to happen.' Consequently, the NUM's task was '*to preserve and to improve in all circumstances, irrespective of what government is in office*, the working and living standards of its members.'[70] In contrast to the NUM's stance before 1951, this represented a potentially significant hardening of the union's position, reflecting the discontent in the NUM. Abe Moffat argued a Conservative government represented a significant shift in the NUM's political and industrial environment, because it had proclaimed its determination to resist public sector pay claims, the NUM's wage claim had to reflect this and its negotiators had to recognise they were dealing with the government. Jim Hammond argued the NUM should use industrial power to boost miners' wages and, rejecting the validity of the wage-price spiral, keep up with price rises; a Yorkshire delegate declared that 'The time has gone by when we are expected to work like Chinese coolies in the mining industry'.[71]

The 1952-53 NUM Debates on Strategy

The debate on the NUM's political strategy had two elements: first, was the strategy of 1945-51 valid and viable under a Conservative government; and second, should the presence of a Conservative government affect the NUM's approach to collective bargaining?

1952 marks the beginning of the debate in the Labour Movement over Bevanism and Consolidationism (and later Revisionism) which was to wrack the movement throughout the 1950s. The NEC's Resolution 34 to the 1952 NUM Conference called for loyalty to the Labour Party and restated the continued vitality of the Attlee government's parliamentary road to socialism and, despite rumblings from the Scottish delegation over foreign policy, this was carried unanimously. Linked was Derbyshire's Resolution 35 condemning the Conservative government. Derbyshire called for a vigorous socialism, including the public ownership of the banks, large engineering enterprises, shipbuilding and the chemical industry, and a disarmament programme. Critics in the NUM argued that Derbyshire's resolution implied that after 1945 Labour had not pursued a vigorous socialist policy and Lawther reminded Conference the issue of party strategy would be resolved at the forthcoming the Labour Party Conference in Morecambe. The mineworkers could not 'afford in any shape or form to question *even to the least possible extent*' the 1945-51 government and if passed, the resolution 'would be welcomed tomorrow as a condemnation of everything that has been done.' The overwhelming need was for unity in the Labour Movement and 'any attempt to split this movement from the policy of the leadership that has done so much and so well for our people would be sheer treason and treachery to the pioneers who built up this movement.'[72] Throughout the 1950s the NUM resolutely supported the Labour leadership and refused to challenge their interpretation of social democracy and the parliamentary road.

Sponsored by the NEC, Resolution 36 noted the Conservative's intention to denationalise steel and road haulage and expressed support for the publicly owned industries and future (unspecified) nationalisations by a Labour government. The resolution warned 'the government that any action aimed to destroy [sic] the unity of the mining industry or to defeat the purposes of Nationalisation will be met by the unanimous opposition of all mineworkers.' Horner accepted the government's declaration that it had no intention of denationalising coal but:

> We know now from experience that a nationalised industry in a capitalist country surrounded by privately owned industry is very largely conditioned by what happens in those other industries, and from a purely selfish point of view it is in our interests that we should keep ourselves surrounded with as many other nationalised industries as possible.

The NUM would co-operate with the TUC to resist any threat to the state sector 'as if it were our own industry which was being threatened.'[73] Resolution 36 was carried unanimously.

The Ministry of Labour's reference back of pay increases proposed by 12 Wage Councils in the distributive trades in July 1952, and the NCB's rejection of the NUM's 1952-53 claim, were attempts to hold the rate of wage increase. Monckton would not interfere with arbitration tribunals (such as the coal industry's NRT) but the Cabinet 'confidently hoped that such tribunals would pay regard in their awards to the economic state of the industry.'[74] The NUM's reaction was that this 'would no doubt have an important bearing on the outcome of their present claim, and in addressing the Board they were presumably also addressing the Government.' The rejection of the NUM's claim convinced the NUM the Board was under ministerial pressure. The NCB stressed it 'had not received any instructions from the Government whatsoever in regard to this claim. Whatever their decision might be, the Board would take their own responsibility for it. Unless and until an increase in the price of coal should be required no approach to the Government should be necessary.' The rejected claim was then referred to the NRT and in October the unprecedented happened, NRT rejected the NUM's claim in its entirety. A shocked NEC noted that 'with the exception of the unhappy experience of 1926 this was the first occasion since national negotiations began that a claim from the [NUM] for a wage increase had been rejected.'[75] In its judgement the NRT emphasised it was not 'bound to conform to the suggestions of wage restraint urged by the Government but the monetary position of the Board and the general economic position require consideration'.[76] Addressing the Special Conference called to discuss the decision Horner described the situation as:

as critical as anything confronting this Union...There are some parallels with the present situation in the events of 1925...when we achieved the pyrrhic victory of Red Friday which was really a cover-up of the then Tory Government preparatory to the destruction which came upon us in 1926...It seems to me that we are deciding for or against a repetition of the events of 1926.

The NRT justified rejection by its acceptance of the NCB's calculation that satisfying the NUM's claim would require a five shilling (25p) per ton increase in the price of coal. Behind the NRT, Horner believed (rightly) was the government so 'We may have to use pressure against this Government, and the kind of pressure is a matter for the most serious consideration by the Executive Committee'. Nevertheless, the NEC did not propose to improve wages by industrial militancy and, faced by the inevitable vociferous protests, it argued 'we think we are at the crossroads...either we are going to find an answer to this impasse as an organised body...or we are going to fail to find an answer, because the answer will not be got in sporadic demonstrations against this Conference.' The Special Conference adjourned to re-open negotiations with the NCB and implied that failure might result in industrial unrest in the pits.[77]

The NEC's response to the NRT judgement was, to the astonishment of the NCB, the submission of a second wage claim seeking a 2s 6d (17.5p) increase for daywagemen.[78] Lawther justified this to the NCB as 'necessary to maintain loyal acceptance of the union's policy in the coalfields where serious dissatisfaction existed because of the tribunal's rejection of the 30s [£1.50] claim.'[79] 'It was a strange procedure', the NCB complained, 'to make a further application before the ink was dry on an Award.' The NCB considered coalfield discontent was due to the NUM submitting wholly unrealistic wage claims and its refusal to recognise the Board's financial difficulties, but the NUM contended that 'the Board should be willing to assist the Union in its dilemma' just as the Union had helped the NCB. The NUM's negotiators argued that discontent was now so severe amongst the daywagemen that it was in the mutual interest of the NUM and NCB to find a solution within the conciliation machinery as 'Neither side wished a position to arise where resort would be made to strength rather than reason.'[80] The NCB agreed but 'Regard must be had to the [economic] position of the industry. It was not possible to take out of the industry more than was put in' and warned the NUM that it was 'not going to be deterred by threats of the use of strength to secure what could not be secured by constitutional means...to succumb to threats, would be the end not only of the [NCB] but of the [NUM] itself.'[81] The NCB was willing to concede an increase of one shilling (5p) on national minima provided wage rates were stabilised until 30 November 1953 except where the average was below 35s (£1.75) per shift in return for an extension of the Extended Working Hours Agreement to 1 May 1953. The NUM and NCB agreed to seek jointly an increase in coal prices to finance an increase of one shilling (5p) in the national minima. The NEC recommended the reconvened Special Conference reject this offer and instead seek local negotiations to improve piece-rates and negotiate a new a New Wage Structure to improve daywage rates.

The NEC recommended the rejection of the NCB's offer and there were immediate calls for industrial action. Using his powers under the 1944 Rulebook, and with the support of the majority on the NEC, Lawther ruled that no other proposals other than the NEC's could be debated at the Special Conference. South Wales' call for industrial action was ruled out of order and despite the anger of some delegations Lawther had the votes to block calls for industrial action.[82] Horner was candid about the situation; 'We have failed. We have not brought the goods home. It is perhaps the first time since nationalisation'. He then outlined the options open to the NUM:

> We can either use our industrial strength, which is for you to consider. It would have to be a national stoppage; it would wreck the whole of our achievement; it would put the whole thing into a position that I do not care to contemplate at the moment. It is either that or making the best use of our opportunities to get the most we can for the day-wage men through these other avenues, and that is what the Executive Committee decided you ought to do.

J. Wood (Scotland) urged the rejection of the NEC's proposed strategy arguing that if 'we are going to have a burst in the coalfield...the quicker we realise it, the better.' The NEC should therefore 'arouse the coalfield to such an extent that the miners will let it be known to the Coal Board and the government that they have got to move in such a situation'. A South Wales delegate reiterated these sentiments arguing 'we could easily make a bonfire of this coalfield if we so desired.'[83] Jim Hammond described the NEC's proposals as a 'grand retreat' and urged the NUM use its industrial strength because 'we are told that this [NUM] of ours is strong. They are always telling us how strong we are. They are always rubbing our muscles, but we sit back and refuse to make use of the strength that we undoubtedly have. We are all conscious of it'. Hammond complained of a lack of national leadership which led to a vicious spat with Lawther, and Bill Allen (Northumberland) agreed the main threat was to the NUM's unity because:

when our members in the coalfields see our national leadership impotent, like paralysed rabbits in front of a snake, when they see you going cap in hand to the Coal Board, and the Coal Board laughing at you, there is danger then of the members beginning to think that the strength has slipped out of the Union.[84]

Alex Moffat also attracted Lawther's fury by claiming that if delegates accepted the NEC's policy 'you will find you have been victims of the greatest confidence trick that has ever been perpetrated.' This prompted Horner to observe that 'when we do not know what to do, to turn and chew each other'. The NEC's policy was approved by 471,000 to 260,000 and was subsequently approved by the Areas.[85] The substantial vote against (35.6 per cent) came from South Wales, Scotland, Northumberland and Cumberland.

NUM strategy was now predicated on the rapid negotiation of a new wages structure and the concession of a second pay claim.[86] The NCB's financial difficulties (and government policy) ensured the Board would drive a very hard bargain. The NUM's difficulty was that the membership expected a pay rise, an expectation which had underpinned acceptance of the NEC's strategy at the Special Conference.[87] The Cabinet discussed the wage negotiations on 20 January 1953. The Board had offered 6s (30p) to 140,000 lower paid mineworkers, whereas the NUM sought 15s (75p) for 420,000 workers. Leathers believed this 'could not be accepted. It was certainly not justified by existing circumstances in the industry where the Tribunal's refusal of the original demand had been followed by a slackening of effort and a loss of output' and the NUM 'would make no effort to check the trouble which would follow in certain coalfields' to increase pressure on the Board and government. Output would be lost but Leathers believed there was no risk 'of any general stoppage of work.' Harold Watkinson, the Parliamentary Secretary at the Ministry of Labour, urged the Cabinet stand firm as 'any further concession by the [NCB] on this occasion might well start a further round of application for wage increases.'[88] Bolstered by this attitude the Board remained adamant that the NUM's full claim would require a significant increase in domestic coal prices at a time when world coal prices were falling, there was growing consumer resistance to high cost fuel, and an increase could not be ignored by government.[89] The emerging compromise was, Leathers told the Cabinet, more than the NCB could afford and risked a rash of wage applications from other groups but the offer 'would encourage the more conservative elements in the Union, who were now prepared to encourage Saturday working and were showing signs of becoming more cooperative generally.' Nevertheless Leathers urged the Cabinet stand firm on its existing offer and sanction no more concessions.[90]

The NEC called another Special Conference for the 27 January 1953 with a recommendation to reject the NCB's offer. If the NUM was ever going to resort to industrial action in the 1950s it would have come as a result of this Special Conference. Horner insisted representatives seriously consider whether they could carry their Areas for industrial action, moreover the level of unrest in the profitable Yorkshire and Midlands coalfields was such that the NUM seemed to be on the verge of a split. He acknowledged that only an increase in the price of coal could provide the necessary money but a press campaign inspired by the government had developed against this. The NEC had been convinced that the negotiations would result in a compromise and were astonished that the standard practice of giving the NUM something had been broken. Horner was:

> honestly of the opinion that after consultation the Board would come back with that offer. I do not know who is the nigger [sic] in the woodpile, but I do know this, that since 1939 the owners and the Coal Board cannot increase the price of coal without the consent of the Government... The Board admits that the only thing they have to consult the Government about is the price of coal. Well, the price of coal is the answer to everything.

The NCB was making a loss and could not meet the NUM's claim without increasing prices but 'the question of prices is a Government matter in which they are helpless, and they have to carry out Government policy.'[91] The NEC therefore recommended by-passing the NCB and approaching ministers directly.

A Lancashire delegate could not understand the NEC's caution because 'if we have to get fourpence increase by going to Churchill, then let us stop Saturday working and let Churchill come to us.'[92] Bert Wynn (an NEC member) understood the rationale of an approach to the government but feared this would be interpreted in the coalfields as 'a complete lowering of our dignity, and there will be disquiet and revolt'.[93] A Scottish delegate saw the NUM's choice as being of wider political significance:

> This is not an isolated fight for a wage increase for miners but an opportunity for the entire working class to challenge the Conservatives. The NUM's policy was inevitably political because if we go down in this fight, that will be the signal for the attacks on the miners' wages and conditions...we have got to use the power we have, and this [NUM] was never so powerful in its life.[94]

A South Wales delegate declared that 'Everyone in this Conference knows that Churchill is the sworn enemy of the working class. Everyone knows that, and that is the man at the head of the Government who will do anything and everything to disintegrate this Union if it is possible.' Bill Kellher (Yorkshire) reported 'there was nothing but cynicism and ridicule from the men in our pit yesterday at the suggestion of going to Churchill...we are becoming messenger boys for the [NCB].'[95]

W.E. Jones, the NUM Vice President, warned an industrial challenge from the NUM 'would be made a political issue, and the miners of this country would again be down the drain.' He accused Communist agitators in the Yorkshire pits of provoking industrial action and 'It is no good talking about wages of our people if, when we have done everything we possibly can to the point of striking to get our own people those wages, we get to a situation where our markets are lost'. If the NUM could not settle with the NCB it had no other option to approach the government.[96] During the lunch adjournment the NEC reviewed the morning's proceedings and in view of the hostility to approaching government the proposal was dropped and an anodyne statement about co-operating with the NCB to improve efficiency substituted. Immediate withdrawal from Saturday working was also rejected on the grounds that many daywage workers relied on Saturday working to boost their incomes. On resumption Horner made a highly significant statement to the delegates:

> there is only one thing left by which to bring real pressure now, and that would be to stop the coalfields. We had better face it now, because some people may start trying to stop them a bit at a time. There will be no stoppage of this coalfield for threepence or fourpence, and there will never be a national stoppage authorised by this Union without a ballot vote of the members before it is undertaken.

The NEC's resolution was approved by 478,000 (62 per cent) to 290,000 (38 per cent).[97] The NUM agreed to extend Saturday working and the NCB increased daywages by one shilling (5p) per shift and the weekly minima by six shillings (30p). Both sides agreed to begin work on the new wages structure and the NCB was especially anxious to lock the NUM into 'a campaign to inform the management and men in the Industry of the facts existing in the Industry, and to emphasise that a happy and prosperous industry and the well-being of the country depended on the creation and maintenance of a true spirit of co-operation between management and men'.[98]

The 1952-53 wage claim is important in the post-war coal politics because it confirmed the NUM's turn away from industrial action which boosted ministerial confidence in dealing with the NUM. Once it became clear that the NUM would not strike ministers sought ways to reduce their dependence on the mineworkers and became more willing to blame the mineworkers for the industry's problems.

Towards a Coal Policy

In private many ministers thought it incredible that the UK was permanently on the verge of a fuel crisis. In working towards a policy the Cabinet agreed that 'It would be dangerous to depart from the principle that the nationalised industries should pay their way...Experience between the wars had shown the dangers of granting any temporary subsidy to the coal mining industry; once the government set foot on that path it would be very difficult to turn back.'[99] 1926 cast a long shadow in government and the NUM. Ministers were asked to consider 'possible means of bringing it home to the miners that, by continually forcing up the price of coal, they were providing industry and households with a sharper incentive to adopt substitutes for coal and thereby increasing the prospects of eventual unemployment in the coalfields.'[100] Leathers and Lloyd told the Cabinet that 'The last thing we could contemplate in our third year of office would be a serious domestic coal shortage'. The Cabinet recorded its 'Serious misgivings' with respect to the level coal supplies which pointed to the 'need for a more vigorous policy in developing our coal resources, and we must consider all practicable measures for that purpose, including measures for increasing the efficiency of the [NCB].' The Cabinet's decision to import coal would be 'a shock to public opinion...and the presentation of this decision would need careful handling' in the light of backbench hostility to the NCB. In a personal minute to the Cabinet, Churchill emphasised the need to find some way to persuade miners to produce more coal and he instructed ministers to reconsider coal policy.[101]

Ministers were actively seeking ways of reducing the country's dependence on coal and improve the NCB's performance. Leathers and Lloyd concluded progress depended on increasing pressure on the NCB. Production was 'a matter which we constantly have under consideration with the [NCB]. We have now had further discussion with Sir Hubert Houldsworth'. They recognised that this might be misinterpreted as ministerial interference with the Board but so urgent was the need for an extra 5 million tons of coal 'We propose to convey to Sir Hubert Houldsworth the paramount need to achieve [this] at least, and indeed to do better'. Even so ministers were uneasy because 'There are so many variables and unknowns that no one can make any useful estimates of output and consumption for more than a year or so ahead'. In these conditions of uncertainty they advocated a shift in policy. Capital investment was showing little return and would not solve the production problem if consumption continued to rise, the NCB could do little to affect the situation so government had a responsibility to promote 'a revolution in the age-long habits of the chief consumers' as well as encouraging a more innovative approach by management.[102] This shift would have to handled carefully and responsibility was given to a Cabinet committee chaired by Butler, the Chancellor of the Exchequer, which began work in October 1953.[103]

GEN 445 was concerned primarily with NCB organisation, production and manpower problems. Despite progress in the NCB's investment programme it had produced little increased output but ministers felt they could not interfere and it should be allowed to continue. However, they agreed that 'The level of investment should be re-examined to ensure that it was adequate [as] investment in this industry [was] of prime importance.' A persistent ministerial complaint concerned the work habits of the mining workforce. Absenteeism was higher than before the war and cost about 10 million tons of coal, the high level of unofficial strikes was also a cause of major disruption. Again ministers felt that government could do little to affect these problems as they were the responsibility of the NUM and NCB and were being dealt with through the industry's consultation and conciliation procedures. Butler was convinced that the price of coal was 'a fundamental element' in the industry and he was convinced that the government had got pricing policy wrong. To discourage wasteful coal use he believed prices should be increased and, in principle, the committee agreed but noted there were 'political difficulties' and that 'an increase in the general price of coal would have a bad effect on industry, and...might raise the pressure for wage increases'. Finally, 'consideration would have to be given to the psychological effect on miners of higher coal prices which were not accompanied by any increase in their wages.'[104]

Despite the huge sums pumped into the industry only one new colliery and 10 major reconstruction projects had been completed. The delay was partly due to the complexity of mining developments but also because the NCB could not take current capacity out of production to speed the reconstruction of existing pits. Major investments were planned at a cost of £200 million but this would 'have no material influence on output in the next few years since 7 years or more must elapse before major schemes begin to pay in terms of output and financial return.'[105] Lloyd doubted that wage increases would boost output as they would have to be paid nationally whereas the need was to improve recruitment and retention of labour in those areas, notably Yorkshire and the East Midlands, with high output potential but where coal was facing serious competition for labour from other industries. A substantial general wage increase might actually result in an overall decline in effort. Lloyd warned his colleagues that the country was on the brink of major fuel crisis and:

> we are likely to find ourselves in 1956 in much the same position of general coal stringency that we are in now, and that we found when we assumed office in 1951. Even this precarious balance depends on a high level of opencast coal, on Saturday working by the miners, on home coal restrictions and the keeping open of uneconomic pits.[106]

The only immediate solution was government encouragement of a substantial increase in the oil burn so 'we must take action where Government pressure or persuasion can be exerted. Public authorities, for example, could use oil for central heating. The nationalised railway and electricity industries could use more oil for steam raising, and the gas industry could make gas direct from fuel oil.'[107]

Lloyd believed that the situation and the proposed solution were politically very dangerous 'because a small margin determined whether demand could be met or not, the country would continue to be at the mercy of the coal industry, and the NCB very much in the hands of the miners'. Oil offered an immediate solution to this dependence but the NUM might resist fuel substitution if they saw it as a direct attack.[108] Butler remained sceptical, doubting that switching to oil would be economic and he preferred to increase the price of coal to discourage consumption, increased prices would also enable the NCB to improve wages in order to encourage recruitment. However, Butler recognised that significant increases in coal prices would provoke the Conservative backbench, produce howls of pain from industry, and give a further twist to the wage-price spiral. He was therefore willing to accept power stations burning oil as long as this was as economic as burning coal. Even if the oil burn was increased Butler believed this would not significantly reduce the NUM's strategic power.[109]

GEN 445 failed to reach a consensus. The issue was put to the Cabinet and both Lloyd and Butler prepared memoranda in support of their positions. Lloyd believed future security would be provided by nuclear power coupled with a 'smaller, highly mechanised and much more efficient coal mining industry' but this would take ten years to achieve. Full employment had altered fundamentally the balance of power in the industry because 'Before the war an extra 50 million tons of coal could be obtained if necessary by drawing on a reserve of 150,000 unemployed miners. Then the spur of unemployment and low wages kept shift output high and absenteeism low. But fear of poverty and of unemployment no longer operates.' Lloyd believed higher wages were unlikely to pull in more labour and so the industry would have to rely on increased productivity and mechanisation but capital investment took 7-10 years to produce coal. The investment programme was predicted to produce some 4 million tons per annum extra but if current demand was maintained this extra output would be easily absorbed without easing the supply problem. The political threat posed by the country's dependence on coal was obvious to all:

> At present the stranglehold of the miners can be absolute. By a three weeks' strike in the winter, which they could well afford they could bring our whole economy to a standstill. By less extreme measures, for example, by a slowing down of Saturday working, they could land us, with the balance so finely adjusted in extreme difficulties.

While the NUM's leaders were willing to continue Saturday working on the basis of annual renewal, the NUM was 'politically opposed to the present Government and would be within their rights in refusing to extend Saturday working'. The answer was more fuel, and only oil could provide the necessary amounts quickly.[110]

Butler did not dissent from Lloyd's assessment of the seriousness of the situation but he disputed the relationship between costs and benefits. The extra 3.5 million tons of coal would cost some £16-£17 million in both imports costs and uneconomic investment. Butler argued:

> These proposals are not in tune with the economic policies which we are following with success in other fields. In contrast to our general policies of freeing markets, we are still holding down the price of coal to a level which does not reflect either its marginal cost or its scarcity. We are not making proper use of the price mechanism.

Lloyd's policy risked open ministerial direction of a state industry which the Conservative Party and government had consistently opposed. Butler would accept measures which did not entail subsidy but believed greater use of the price mechanism would produce the desired outcome. He acknowledged his policy would cause trouble with a section of the backbench, domestic consumers and industry, and he agreed his approach did not address the problem of dependence on the NUM. Indeed, his advocacy of 'continuous and energetic recruiting' until there were 720,000 miners on the colliery books threatened to increase that dependence.[111]

Fearing Butler would prevail in Cabinet, Lloyd wrote directly to Churchill stating 'we are both convinced that it would be wise to bring in a few million tons of fuel oil to supplement our coal supplies. This would help loosen the stranglehold that the miners now have on our economy'. The Cabinet Secretary, Sir Norman Brook, thought GEN 445 was 'not very satisfactory, but I doubt whether any good will come of referring it back to the Committee', but concluded the Cabinet would have to settle the policy. Three issues had to be resolved; the first 'is whether they endorse in principle the plan to increase our domestic use of oil fuel as a competitor with coal'; second, 'whether the other nationalised industries can properly be asked to undertake, in furtherance of a national fuel policy, developments which they would not regard as commercially sound'; and finally, the Cabinet would have to determine the balance between expanded open-cast mining and the demands of agriculture. The first two of these issues went to the heart of the Conservative government's management of state capitalism. On 26 May the Cabinet agreed to postpone a decision and encouraged Lloyd and Butler to come to an agreed position.[112] The difference of opinion between Butler and Lloyd remained the main obstacle to an agreed policy but all agreed the central problem was 'that the miners will not produce all the coal we need'. Production had to be increased and consumption cut; the debate was over how. Lloyd wanted to burn oil, Butler wanted to use economic incentives but the Cabinet Secretary believed the Chancellor would not press his views as it was unlikely he could persuade the Cabinet 'that it would be politically wise to raise coal prices further'. This would avoid a hostile reaction from the Conservative backbench and industry, and Norman Brook believed Lloyd would be able to persuade the British Electricity Authority and British Railways to burn oil without public direction. He advised Churchill that the Cabinet could be persuaded to approve an increased oil burn.[113]

Cabinet resumed its discussion on 2 June. Ministers needed no persuading about the need to reduce dependence on the NUM but they did worry about the political impact of open cast mining on the farming lobby and the experimental burning of oil at the Marchwood power station in Southampton on a public opinion increasingly sensitive to air pollution. Oil could be supplied to Marchwood and other estuary power stations 'at prices so favourable that the British Electricity Authority were prepared to entertain the possibility of installing alternative oil-firing as a sound commercial proposition' and the option to return to coal burning would be retained. This encouraged Butler to withdraw his objections and the Cabinet approved the policy of oil substitution.[114]

This decision did not transform instantly the situation as a shortfall of even a few million tons of coal would still have serious economic and political consequences.[115] Lord Woolton reflected a view in government and Parliament that it was ludicrous that despite 'the abundance of our resources of deep-mined coal' the country lived in perpetual fear of a fuel crisis. Since 1947 wages had increased substantially and a huge amount of capital had been invested in the pits but 'there are still no signs of material progress towards either sufficient or cheaper coal', moreover steadily increasing coal prices undermined competitiveness which 'endangers full employment and hampers our efforts to reduce the cost of living.' Woolton concluded that 'We cannot afford to be at the mercy of the present monopoly, which threatens both our standard of living and our trade' and the solution was 'that it would be wise for the State to make a deliberate effort to encourage the burning of oil in competition with oil.' Jock Colville, Churchill's secretary, brought this to the PM's attention who 'noted' Woolton's suggestions.[116] Woolton's suggestion that the government use the oil companies to mount a frontal assault on both the NCB and the NUM was too radical a solution for Churchill. However, the narrow safety margin was brought home sharply to the Cabinet in December 1954 when, suddenly, the government found itself on the verge of a fuel crisis. Lloyd warned Churchill that coal stocks were too low for comfort and although remedial measures had been put in place a possible railway strike would 'bring most of the pits to a standstill very quickly because they cannot stock more than a small quantity of coal at the pithead.' The Cabinet were warned 'coal output would soon fall by about 3 and a half million tons and this would make it impossible to get through the winter without a fuel crises' [sic].[117]

In April Lloyd submitted a paper to the Cabinet's Economic Policy Committee which set out his proposed response to the problem posed by the NCB. Part of this involved a more professional managerial ethos and elite. After a mini-purge a new Coal Board had been appointed by Lloyd in January and had embraced the Fleck Report on the organisational reform of the NCB and which was 'genuinely determined to strengthen management throughout the industry'. The second problem (the NUM) was more intractable and there was no immediately apparent solution because:

> there is no doubt that one of the chief reasons for poor discipline in the pits in the dominant influence of piece-workers at the face, and a major cause of their undue power is the post-war scarcity of coal. Their scarcity value can be weakened only by reducing the immediate urgency of the need for increased output, and this can only be done by economising in the use of coal by greater use of oil in its place.

Nuclear power could only make a small contribution to national energy needs by the early 1960s and he 'was satisfied that until then it will be necessary by all possible means to make oil increasingly available as a substitute for coal.'[118]

In Cabinet 'There was general agreement that the coal industry was failing in its proper part in the national economy. Higher coal prices were not an acceptable substitute for increased production and greater efficiency'. The Fleck Report, and a request the Board reconsider any proposals for price increases, were designed to focus managerial attention on productive efficiency.[119] The July 1955 Commons debate was widely expected to produce a rebellion from Conservative backbenches. The debate provided useful publicity for the oil substitution policy as well as airing criticism of the industry for its poor performance. It was hoped that this, when coupled with the NUM's refusal to accept foreign labour in the mines, would force the NUM onto the defensive.[120] Lloyd emphasised the continued parlous state of coal supplies which would only be resolved when reconstruction was completed and the government's oil policy was presented as a temporary expedient to provide coal with a breathing space. Citing his wartime experience as petroleum minister, Lloyd proposed the government and country should take advantage of 'the immense flexibility and the enormous productive power of the oil industry.' Political stability in Iran (largely as a result of an Anglo-American coup against Iranian nationalists) had led to an increased flow of oil, the coal price increase of May 1954, and the Marchwood decision had encouraged coal substitution so 'that the spectre of the "coal gap" is removed.'[121] Nabarro repeated his long-standing criticism of government policy and the NCB, but supported Lloyd's policy as 'it is better to have a competent Conservative Administration in office, with a lapse of policy in the Ministry of Fuel and

Power, rather than risk the return of a wholly incompetent Socialist Administration.'[122]

High coal prices would encourage oil usage and as oil replaced up to three times its own weight of coal any coal price increase would have a multiplier effect making further oil conversions attractive.[123] Even so ministers were advised that oil conversion would take two to three years to become effective, so coal would have to be imported to guarantee a safety margin. The coal supply/use margin was extremely sensitive to the weather, a cold spell could mean the difference between power cuts and keeping the lights on. Lloyd warned the Cabinet that 'Until the middle of February the winter was on the whole a mild one...Despite this mildness our coal supplies position has throughout been delicately balanced and total distributed stocks have now fallen to a dangerous level.' The Ministry of Fuel had been compelled to direct supplies to essential users and had imported coal and yet it was predicting a coal gap of 7.8 million tons by the end of the summer so its objective was to create a stock of 19.8 million tons supplemented by imports, a total stock of 22 million tons. Railway coal stocks were low, there were coke supply problems in the steel industry, and the electricity and gas producers were to be required to economise further in their use of coal but the oil companies could supply an extra 4.5 million tons of heavy fuel to the power stations and by 1958 oil would have replaced 7.5 million tons of coal. If coal output did not improve Eden was warned there would be a need for emergency measures to concentrate the remaining coal stocks at the power stations to avoid 'our drifting into a 1947 fuel crisis.'[124] Not surprisingly there was great frustration in government with the coal industry.

The Fleck Report

The Churchill government would have liked a wide-ranging inquiry into the NCB similar to one conducted into the electricity supply industry 'but were fearful of the [NUM's] reaction, and instead left the Coal Board to arrange its own enquiry.'[125] Whilst correct this does not convey the extent of the government's, or its backbenchers', disquiet with the NCB's structure and the quality of its management. The Conservative government increased significantly the extent of ministerial involvement in the industry's affairs whilst resolutely maintaining the fiction of non-interference. Ministers resolved to reform the Board's structure and culture by introducing decentralisation despite the danger inherent in such a policy. Afflicted by a constant fear of a fuel crisis the Cabinet agreed that something had to be done about the NCB but the Cabinet felt it could not insist publicly on a review of the industry's management.[126] The NUM was similarly inhibited. The NEC feared criticism of the NCB by its members might be used to open the door to decentralisation and this seemed confirmed when Conservative MPs called for an inquiry into the NCB.[127] Two months later an NEC Sub-Committee examining the NCB's structure concluded that 'On the evidence before us we are of the opinion that, in general, there is no reasonable cause for dissatisfaction with the Establishment of the [NCB].' The NCB indicated its willingness to investigate specific complaints from the NUM's Areas but was not interested in generalities.[128]

Lawther used his final Presidential Address in 1953 to call for an end to attacks on the NCB and public ownership of the coal industry. 'This Union', he pointed out, 'has never argued that the inauguration of nationalisation was the end of our troubles, nor do we believe the [NCB] is the perfect form of organisation'. For the mineworkers to prosper 'We must get this industry away from the stage of hanging onto grievances both ancient and modern' and nationalisation had to be defended resolutely from political interference.[129] Nottinghamshire called for a full inquiry into the growth of bureaucracy and the absence of working class representatives on the Board but came under great pressure from the NEC to withdraw its resolution which it agreed to do. Welcoming this decision, Jones argued the resolution contained features which 'if they had been debated here, would have given rise to misunderstandings in the public press, and they could also have been misinterpreted in those spheres of public life that are contrary and in opposition to the Labour and trade union movement.'[130] Nottingham submitted a resolution to the 1954 Conference expressing concern at the changing nature of the NCB. In particular it criticised its inclination 'to put the University man before the Technical man', and complained about the decline in the influence of the Labour Department within the NCB. The resolution's mover argued 'men who have spent a lifetime in the industry, men who were and still are staunch trade unionists, and in many cases men who have helped to create nationalisation' were being replaced by University trained personnel officers who were 'the forerunners of a new regime' and soon there would be no room in the NCB 'for the ordinary man at the pit.' The NEC asked for remission on the grounds that it did not wish to discourage anyone who could be of service to the industry.[131]

Disquiet with the NCB's management skills amongst MPs and ministers required a political response.[132] Ministers were vulnerable 'to much criticism for their failure to do something to clean up the mess inside the Coal Board.' Many backbenchers agreed with Brendan Bracken's assessment of the NCB as 'a crazy contraption', 'a rotten organisation' and described the chairman, Houldsworth, as 'a pleasant second rate lawyer with Left Wing affiliations. He has never had any experience of business, nor has he any force of character. Two of his principal colleagues are ex Trade Union Officials, basking in their free motor cars and their £5,000 a year salaries. They are utterly useless as controllers.'[133] Throughout the spring and autumn of 1953 Leathers and Lloyd had held frequent discussions with Houldsworth on the coal problem and these talks convinced them of the need for organisational change at the NCB. This raised a major political problem for 'if the changes appeared to the miners to have been influenced by the Government there is, as some of the speeches at the Miners' Conference showed, the danger of a violent reaction which might lose us a great deal of coal and land us in severe difficulties. The timing and handling of this matter must be left to the [NCB].' Houldsworth agreed on the need for organisational reform but was 'having no easy task to convince some of his colleagues on the Board'.[134]

Organisational change was regarded as so vital for the industry's future that extreme political caution was required so 'not to arouse the opposition of the [NUM]' and ministers were warned 'that nothing should be said that might convey the impression that they result from Government pressure.'[135] The Cabinet were in no doubt as to the sensitivity of what they proposed and were anxious not give the impression of surrendering to backbench atavism because to do so 'would impair the good relations which had been established between management and men in the industry.' Lloyd knew that some MPs would regard an NCB sponsored inquiry as insufficient but he 'hoped that Government supporters would be content with these measures, which were *more likely to be acceptable to the miners* than any enquiry undertaken at the initiative of the Government.'[136]

The decentralisation of authority to NCB Area General Managers was 'a major measure' that would limit contact with the national Board allowing the Board to concentrate on policy and the Areas on output. Of greater significance was the NCB's appointment of what would be the Fleck Committee which would have 'to make progress without antagonising the miners'. This would be difficult as the ministerial and Board consensus in favour of decentralisation was bound to arouse NUM suspicions, but 'The coal industry had to be got back to a system under which a suitable production unit had adequate responsibility, including control over its own recruiting and promotion policies, and which would enable morale, and with it production, to be brought to a high level.'[137]

The organisational evolution of the NCB since 1947 and the coal supply problem had serious implications for ministerial relations with the NCB. The specific cause of controversy was the esoteric issue of briquetting. Despite ministerial prompting the NCB was reluctant to invest in briquetting to provide the railways with an alternative to the large coal they traditionally preferred. The NCB which was responsible for the efficient operation of the industry, regarded this as an unproductive investment but under CINA ministers had a duty to safeguard the national interest. Lloyd noted that 'In the past the Board have often followed my advice on hearing informally what I believed the national interest to require. On this occasion I may need to give a formal direction and that I would propose if necessary to do.'[138] To do so would be a radical departure from past practice and politically dangerous as it would confirm the NUM's belief that government was pulling the NCB's strings. Lloyd was dangerously close to publicly demonstrating the real extent of government control over the NCB. In a Cabinet paper of May 1954 he wrote 'We must continue the most determined efforts to secure by better leadership, labour relations and organisation a change of spirit in the industry. *I am in constant touch with Sir Hubert Houldsworth...on these matters'.*[139]

The likelihood of an adverse reaction from the NUM to Lloyd's enthusiasm for organisational and cultural engineering worried some of his Cabinet colleagues. Butler was unhappy with a ministerial direction instructing the Board to undertake capital investment which managers deemed uneconomic. Ministers could only contemplate this action because coal was publicly owned, 'But one of our strongest arguments against nationalisation has been that it would lead to just this type of undesirable interference by the Government.' To do so would do the government serious damage amongst its supporters and in the country.[140] Churchill was warned that this was a major issue, pregnant with dangers for the government because 'The question here is whether the national interest should be allowed to override the principle that nationalised industries should run their businesses on normal commercial lines.'[141] The problem was how to direct the NCB without making these directions formal and therefore public. By 1953 there was an acceptance in government and on the Board that the division of labour between the national Board and the Areas was messy. This resulted in the General Directive (October 1953) which codified much of the practice which had grown up since 1947 and owed much to Houldsworth's conception of what was effective management. The General Directive was received with hostility at NCB headquarters because it reduced the centre's influence and enhanced Houldsworth's 'presidential' style which had alienated many of his colleagues.[142] Houldsworth's abrasive style led to in-fighting in the NCB and management style was an issue addressed by the Fleck Committee appointed by the NCB in December 1953.

Chaired by Dr Fleck, the chairman of ICI, the committee worked for twelve months taking evidence from all major interests in the coal industry. It acknowledged the soundness of the NCB's basic structure but urged uniform administration, that full-time Board members should have knowledge and experience of the industry and be responsible for a specific aspect of the NCB's operations. Knowledge and experience of the industry were to be sole the criteria of appointment to divisional boards. In other words, Fleck pushed the technocratic management ethos inherent in the Morrisonian public corporation further. Fleck's other recommendations related to the general quality of NCB management. The poor quality of management was one of the inheritances of the pre-nationalisation industry but Fleck was also critical of the philosophy underlying the General Directive and, therefore, of Houldsworth's management style and philosophy.[143] Fleck concluded that 'The NCB had too few managerial staff and their average quality was low. Therefore they ought to get rid of the worst, increase the staff and train everyone to a better standard.'[144] Fleck stressed the effect of the complex historical legacy of the coal industry, which was not widely understood and which had led to much ill-informed criticism, and his report argued that since 1947 the industry had done a remarkable job for the nation.[145] In this period the NCB had been subject to contradictory demands (maximise output whilst modernising the industry) and this confusion had to be ended by giving the Board a clear strategic objective. This was needed because 'The Board cannot make a profit in the sense of building up a surplus from which to distribute dividends to private shareholders....In short, the British coal industry is now a public service.' Fleck noted there was nothing in the nationalisation act 'that prevents the industry being managed in accordance with the best commercial practice. Nor is there any conflict between a commercial approach on the Board's part and the Board's being a good employer.'[146] Despite public ownership there was no reason why the NCB should operate any differently from Fleck's own company, ICI.

The NUM remained concerned and warned that any diminution the central NCB's authority over planning, production and finance would be resisted by the NUM. Its 'unqualified opposition' to decentralisation was based on the rationale that public ownership was intended to treat the industry as a single entity which should be developed in a coherent way. The NUM 'as a result of bitter experience, recognises the importance of maintaining national negotiations with respect to wages and working conditions, and is aware of the danger of a return to the settlement of these questions on a coalfield basis.'[147] The NUM did not object to refining the NCB's organisation but it recognised this entailed its exclusion as the management strata was professionalised; 'it is vital to the success of the mining industry...that measures be taken to utilise to the full the accumulated experience and ability that is crystallised in the [NUM].'[148] Much of this experience could be deployed in the Board's Labour Relations Departments which the NUM believed should be staffed from the labour side. However, 'From every coalfield the reports suggest that there has been developing a tendency to subordinate the Labour Relations Departments to other Departments of the board, in particular, to Production and Finance.' NUM policy was to 'press that all those appointed to positions in the Labour Relations Department of the [NCB] shall be selected from the trade union membership within the mining industry.'[149] Compare this with Fleck's conclusion that 'the Board would be ill-advised to rely as heavily as they have done in the past on the ranks of the trade union officials as a source of recruitment' to the Labour and Welfare departments. All the then Divisional Labour directors had been NUM officials some coming into management as a result of wartime pressure, others as a result of nationalisation. The Board was recommended to draw on a wider range of personnel.[150]

Fleck was accepted by the government. The absence of an adverse reaction from the NUM emboldened ministers and the new structure was speedily established. In February 1955 all the full-time Board members submitted their resignations to Lloyd (two-part time members also resigned and were not replaced, the remaining two leaving at the end of 1955) and only two of the pre-February Board's full-time members were re-appointed; five new members were appointed, including James Bowman as Deputy Chairman (Ebby Edwards had left the Board in October 1953), and when Houldsworth died in February 1956 Bowman replaced him as Chairman. The new Board enthusiastically embraced Fleck's agenda and ethos, and several members of the Board had been keen to move down this path for some time but were hindered by hostility from the NUM and a lack of political support. The NCB was 'genuinely determined to strengthen management throughout the industry' and work began to create a department 'which will be concerned with the recruitment and training and managers at all levels.'[151] Fleck also brought a much higher level of ministerial tutelage of the Board because 'We must seek to secure our reorganisation of the [NCB] leads to fundamental improvements in the industry. Quick results in terms of increased output cannot be expected; but technical development and reconstruction underground must be matched by a managerial renaissance and better discipline throughout the industry.'[152] The use of 'our' says much about the inspiration for these changes and shows ministers were now willing to take a more direct role in the industry's affairs, the Board now accepted the minister's right to issue 'informal-formal' directions, and the NUM had shown no inclination to challenge the new policy. By the summer of 1957 the post-Fleck Board was established and was 'very different from that of the Houldsworth era; stronger at the top...with more professional knowledge and more harmony.'[153] Fleck recommended that this should be the last review into the industry for some time and that it should be allowed to settle down. Ministers agreed.

The NUM's response to Fleck was one of resigned acceptance.[154] Horner stressed to Conference delegates that Fleck was concerned with the NCB's organisation not nationalisation *per se* but warned delegates that criticism of the NCB would be reported in the press and would encourage those hostile to public ownership to push their hostility further. He took the standard NUM line that whatever the NCB's weaknesses the mineworkers' position was 'enormously better than it was when the industry was privately owned.' Those who had worked under both production regimes knew which they preferred, the Board had striven to work constructively with the NUM and any problems the unions experienced 'are in the main problems of growth, and not the problems of decline.'[155] Delegates agreed that the NCB was better than the coal-owners but also argued that the NUM had a right and duty to be critical. A Lancashire delegate reiterated that substantial numbers of managers remained hostile to the conciliation machinery believing it 'must be used exclusively by the men and men alone.'[156] A resolution criticising NCB bureaucracy and protesting at absence of manual workers at the highest level of management was submitted by Nottinghamshire. This resolution was carried with minimal debate.[157]

In December 1956 the NUM was asked by the minister to make nominations for the post of NCB Industrial Relations Director. The Midlands Area urged the NEC to argue that appointments in Industrial Relations Departments should 'at all times be filled from the [NUM]'.

Although appointments were the NCB's responsibility the NEC felt the NUM had a vital interest in Industrial Relations appointments and the Board were urged to consult widely on such appointments.[158] Responding to rumours that a non-miner would be appointed the NUM's National Officials sought a meeting with the Minister of Fuel and Power. The NUM's case was that it was vital for workforce confidence in management that the IR Director have a mining and union background. The Minister listened politely and appointed Sir James Crawford who was not only a non-miner but came from outside the coal industry, although he had been president of the Boot and Shoemakers' Union. The NEC were incandescent with rage and cautioned that the appointment 'had caused serious repercussions in the coalfield and the strongest possible resentment was being expressed by the miners. Areas were being faced with resolutions from Branches advocating action ranging from a general stoppage, withdrawal from Consultative Committee at all levels, to go-slows and withdrawal of Saturday labour.'[159] The minister was condemned and Crawford was called on to resign. This appointment was symbolic of the NCB's evolution in a direction not liked by the NUM but the minister's justification could have come from Herbert Morrison: 'it was his responsibility to choose the man who in his opinion was best qualified' and Crawford refused to resign. Some NEC members argued the NUM should show is displeasure by withdrawing from the conciliation machinery but, 'despite the very strong resentment which was felt at the action of the Minister, it would not be in the best interests of the membership to take any direct action at this stage.' Or at any stage. The NEC noted that 'in the whole of the present [IR Department] at the [NCB] *there was not a single person who had come from the Union.*'[160]

NEC proceedings and Conference debates reveal a steadily declining interest in appointments indicating a recognition that a threshold had been crossed. NUM – NCB relations had settled into a pattern based on traditional and bureaucratised management-union roles and a familiar collective bargaining agenda. The presence of a Conservative government after 1951 inhibited the NUM and muted its criticism of the NCB, and the NUM was very reluctant to acknowledge there existed deep-seated discontent with public ownership and the NCB amongst the mineworkers. The Board recognised the NUM's 'special interest' in senior appointments but 'were unable to accept that all appointments...should at all times be filled by men from the ranks of the Union.' The Board reassured the NUM that whoever was appointed 'would always have the interests of the industry and the mineworkers constantly at heart.'[161]

Sam Watson feared the industry's management was becoming dominated by 'men who have little or no experience of the industry' and within a decade the men who had brought about public ownership would have retired. If the industry was going to be kept out of the hands of the 'managerial class' the NUM would have to become more closely involved in administration. A.J. Pratt (Midlands) believed that the NUM had made a serious error of judgement when it had refused to become involved in Board appointments in 1946, and Jim McKendrick (Scotland) quoted Bowman's injunction to the 1956 Conference ('We march together'), adding 'that the hangman can say that to his victim.' Even when NUM members joined the NCB the union received little in return. When James Bowman joined the NCB 'we also expected that his loyalty was still with us and we look for material evidence of that.' Bert Wynn (Derbyshire) regretted the NUM had not done more to challenge these developments in what was their industry. They did not do so because the NUM believed that nationalisation was the end rather than the start of the battle and so 'we had no policy in regard to the evolution of this industry...We have stood still for ten years, and there has been no thought applied to this very important problem.'[162] Horner reminded delegates that when the industry was nationalised in 1947 Bowman, Lawther and himself concluded that in a market economy a publicly owned industry would be severely circumscribed and so the NUM had to remain a free and independent union to defend its members, but also participate fully in the industry's development. Public ownership was not socialism and there remained a 'fundamental contradiction' between the NCB as the buyer of labour power and the NUM as seller which would remain until socialism was realised. As NCB chairman Bowman could not be expected to 'occupy exactly the same role' as when he was an NUM official as his job was to safeguard the industry, but his transfer to the Board did not automatically make him the NUM's enemy. Horner admitted that NUM had 'a sort of Dr Jekyll and Mr Hyde' existence but significant gains had been made and the industry was still developing so all was not lost.[163]

Avoiding Power Politics

After the 1952-53 crisis the pattern of Board, NUM, government relations was that as long as the *forms* – consultation plus the incremental improvement of pay and conditions – were obeyed the NUM would not revert to pre-1947 attitudes or sanction the revival of the class struggle in the pits. This was the political danger posed by the NRT's Twenty-Fourth Award rejecting the 1951 wage claim because it seemed to presage a recasting of the rules of the game and explains why the NCB conceded the second claim in order to stabilise the conflict management system in the coal industry. Without a concession the NUM's role in managing discontent in the pits, a role already complicated by representing its members interests to management as well as sustaining nationalisation, would have been compromised. These contradictory roles were reconciled by a variety of strategies, such as reference to 1926 and defending public ownership (whatever its faults) against the depredations of Conservative backwoodsmen. This reconciliation was dynamic, indeed unstable, and had to be renegotiated at the NUM Conference, but it remained in place. In maintaining this equilibrium the NUM had an important ally in the Conservative Government. No attempt was made to establish a specifically Conservative model of public industry, ministers preferred to strengthen covertly NCB resistance as a substitute for closer direct control of the industry, and the NCB's growing deficit encouraged indirect control. The Conservatives had no political incentive to change radically the relationships created under Labour and as long as the forms were obeyed the NUM was content with the status quo and ministers accepted a high but unofficial level of unrest in the pits.

This balance was sometimes disturbed. 1953 saw an upsurge in wider industrial unrest and there was speculation about when (not if) the miners would be drawn into this wider conflict. Abe Moffat argued the NUM should be part of the wider union offensive and 'So long as the Government continues to carry out that sort of policy, it is absolutely imperative that we as a trade union take the necessary steps to protect our members.'[164] Sam Watson strongly opposed this; 'I would sooner be thrown out of my job...than make the rank and file believe that we can go forward on a policy that we cannot substantiate.'[165] A South Wales delegate warned that the low paid were 'no longer going to accept the position whereby [they] are going to be the poor relations of a prosperous industry...The time has come, if we cannot have it from the [NCB], then we have to re-assess our sense of values within the Organisation.'[166] Industrial action would represent a monumental shift in the NUM's stance and Horner argued 'This mechanical attempt to get increases *in the absence of mass agitation and mass pressure* will not work any more. We have passed into a new sphere...we have to build up a propaganda machine to demonstrate the justice of what we are asking for'. Horner refused to discuss these new tactics for the simple reason that 'We just have not worked out the means to realise our aims'. The resolution was remitted.[167] The NUM engaged in neither mass agitation nor pressure.

The NCB rejected the 1953 claim as ill timed and expensive. The 1946 conciliation agreement dictated the claim would go to the NRT but the Twenty-Fourth Award had shaken severely the NUM's confidence in the procedure. The one response that was not seriously considered was industrial action so the NUM would to go back to the NCB and 'take a stronger line' short of industrial action, such as ending the EHA. Even this modest threat was too much for some Executive members who feared this would lead inevitably to the destabilisation of the conciliation machinery.[168] The NUM reminded the NCB that 'there had been no threats made by any responsible members of the Union' and production had improved. At a time when the NEC was subject to growing pressure from all coalfields it was essential, the NEC argued, that 'continued confidence should be maintained in the leadership of the Union' because any weakening of its legitimacy and authority had serious political and industrial implications:

> If discontent with their leadership were to lead to the weakening of the Union this would be disastrous to the industry. The Union did not want to talk "power politics" but they thought it as well to make it clear that the Executive knew well enough the strength of the mineworkers position. There were certain sanctions within the power of the mineworkers which if exercised could do great damage. They looked to the Board, to make possible the continuance of the avoidance of power politics by the Executive Committee.

There was a 'universal feeling that the [NEC] had misled the men, and that if the men had taken [strike] action on the lines of that taken recently in certain other industries they would have had a satisfactory result to their wage claim long before the present time'. The NCB recognised the NUM's difficulties and the Board denied government interference when it increased day wages for the sake of industrial peace.[169] Government may not have 'interfered' but the NEC knew what was going on. Houldsworth briefed both Lloyd and Monckton who in turn briefed the PM about progress, which was described as 'relatively favourable', with the NUM's claim. Houldsworth believed the NUM would accept £7 10s (£7.50) as this was comparable with the earlier railway settlement and would be sufficient to avert any threat to the 12m tons of coal produced by Saturday working. This would increase pithead prices by 1s 6d (7.5p) a ton but Lloyd and Monckton saw this as a price worth paying because of 'the importance of reaching an agreement that will strengthen the right wing men who are helping so much, and who have to carry the election of the Union President next month against the Communist, Abe Moffat.'[170]

The sense of crisis in the NUM was reinforced by the Conservative victory in the 1955 General Election. The Conservative manifesto advocated nuclear power and oil as substitutes for coal prompting the NEC to warn that 'any Tory Government that tries to take us away back thirty years and to thrash us as they did at the colliery office doors when seeking a job – that time has gone, and if they do not realise it, then it will be God help this country.'[171] Some members of the government were now prepared to think the unthinkable. In September 1955 an unidentified Cabinet minister urged on Lord Woolton the need to secure 'some contraction in the demands' being made by the unions. The NUM's refusal to accept foreign labour into the pits was cited as an unreasonable exercise of union power and suggested a new approach to the miners:

> I believe it would pay us to get really tough with them. They have dominated the situation for much too long, and if they were to threaten a strike it would have such a calamitous effect on every other section of the public that I believe the TUC would use all its influence to persuade them to think of the interests of the nation.

Woolton passed this to Eden but as '[Eden] had just been down a mine in Durham and was so impressed by the cordiality of the reception that he had received that he could not bear to contemplate anything so rough.'[172]

The potential for unrest in the pits was demonstrated in a major strike in the Yorkshire coalfield over anomalies in the 1955 wage restructuring. Eden was concerned over the strike and asked to be kept informed. He noted approvingly that both management and the NUM were standing firm against the strikes but Monckton attached no great significance to the strike seeing as typical of the volatility of the Yorkshire coalfield.[173] Disquiet was not confined to Yorkshire. At the 1955 Conference a Nottingham delegate warned the Executive 'do be careful how you tread, or they [pieceworkers] will not be as quiet as our day-wage workers have been.' Dai Francis (South Wales) warned 'the men are very restive. A strike at one pit on any one of these issues would spread like wildfire.'[174] Three issues – hours, wages and industrial action – dominated NUM politics. Working hours were of enormous symbolism as the miners had lost the seven hour day as a result of their defeat in 1926 and although restoration had been included in the 1946 Miners' Charter no progress had been made and hours had been voluntarily extended. The claim for a seven hour day underground and a 40 hour week for surface workers was potentially explosive and because of the symbolism of the hours question the NEC accepted the resolutions which were passed unanimously.[175] Nothing was done to reduce surface hours and this began a process which was to culminate the massive unofficial strike wave of October 1969 protesting at the NUM's failure after fourteen years, to secure a forty hour week inclusive of mealtimes for surface workers.

Wage pressure in mining continued despite the government's policy of restraint. Harold Macmillan, the Chancellor of the Exchequer, urged that ministers responsible for nationalised industries bring pressure to bear on their chairman not to capitulate to pay claims, particularly on the railways and in the mines.[176] Paynter warned of the 'tendency for some to live in a fool's paradise. We have full employment; there is an underproduction of coal in the country; the consumption of coal increases more rapidly than we can produce it, and there is sometimes a tendency to believe that that state of affairs will go on for ever'. He concluded presciently, 'we have to make hay while the sun shines'. A.J. Pratt informed delegates 'We have just recently had a strike [over day wages] at one of our collieries. It had to be an unofficial strike but the men were right, you know'. A miner could earn £8 7s (£8.35p) for a forty hour week in the pit compared to £10 at Courtaulds. Why then should they stay in the pit?[177] This discontent was reflected in the Yorkshire Area's attempt to change the 1944 Rulebook to institute the re-election of the National Officials after five years in office. Yorkshire's resolution argued the gap between leaders and led could be bridged by periodic election and any official representing the men's true interests had nothing to fear from re-election. Yorkshire was supported by Nottingham, Scotland and South Wales. As always the NEC's recommendation was to reject as change might result in the removal of experienced officials forced to undertake unpopular measures for the wider good of the union and industry. Political campaigns would inevitably be organised, officials would waste time campaigning, factions would develop and the union's effectiveness would decline. This was the classic defence of oligarchy.[178] A rule change required a two-thirds majority so it stood no chance but the narrowness of the vote, 353,000 in favour to 369,000 against, was an indicator of disquiet.

NUM – NCB relations remained difficult. The NCB, for example, wanted to use foreign labour to ease manpower shortages, a solution rejected by the NUM. Instead the NUM sought a second Miners Charter of shorter hours, three weeks paid holiday, full pay for six weeks of illness per year, plus a general wage increase to boost recruitment and ease tensions in the coalfields. In response to the NCB's 'disappointment' with this claim the NUM issued a ritual warning that 'The union was aware that the use of its industrial strength might very well result in serious consequences for the nation as a whole and it was essential that the negotiating machinery worked efficiently.'[179] Aubrey Jones, Lloyd's successor as minister, took these threats seriously enough to warn the Cabinet that '[I]n view of the very strong bargaining position held by the miners some concession is necessary. I am discussing with the Board the possibility of some kind of "Treaty" with the union under which concessions on the miners' charter would be made only in return for advances in practice'. Jones recognised the political delicacy of this proposal and emphasised there was no guarantee that the NUM could deliver its members support for a 'treaty.'[180]

The government were uncomfortably dependent on the NCB's firmness and negotiating skills. Ministers observed closely negotiations and influenced them by 'an off-the-record talk' with the Chairman or Deputy Chairman whose assessment was based on the informal sounding of the NUM President and Vice-President. The NCB chairman, James Bowman told Aubrey Jones that 'after sounding Ernest Jones and Arthur Horner' (his old colleagues at the NUM) the Board would offer 2s 3d per shift and Bowman made it clear to Jones that if the government wanted peace in the pits an increase of this order (7.5 per cent to 8 per cent) was required.[181]

By 1955-56 the relationships fundamental to the post-1947 coal industry were under strain.[182] The NCB restricted the increase to 7.5 per cent and secured the continuation of Saturday working without conceding any of the NUM's other demands as well as securing a commitment from the NUM (in cooperation with the NCB) to make a 'determined attack' on restrictive practices, absenteeism and unofficial strikes. This was regarded as a major triumph for Bowman and led Eden to comment that 'Bowman has done very well.'[183] It occurred 'at a time when the industry was asking Parliament, during a credit squeeze, for authority to borrow £400 million over the next five years. It had to do so shortly after announcing an accumulated deficit up to the end of 1955 of £37 million'. The NUM was warned that coal was failing to provide the fuel the economy needed and it was facing increasing competition from oil and atomic power, developments which threatened their members' standard of living.[184]

Politicians, civil servants and some union leaders were beginning to question the sustainability of the post-war settlement and the interconnected problems of public spending, full employment, and union power were coming to dominate the political agenda.[185] The Conservative government was losing patience with the coal industry. Over lunch at

Chequers Eden told Houldsworth and Bowman of his intention to protect the NCB's investment programme and 'he hoped the industry would show their appreciation of this by increased output', but in response Houldsworth only '*hoped* to see some results from their capital expenditure by the end of 1957'. An exasperated Eden 'pointed out it was absurd to buy Belgian coal, largely mined by Italian workers, when foreign miners could produce coal in this country; if this position continued the country would be annoyed with the miners.' Houldsworth and Bowman glumly agreed but could offer no suggestions.[186] In 1955 coal costing £80m had been imported to insure against the effects of a severe winter because of 'disappointing colliery output' but plans were laid 'that take advantage of the experience gained during the railway strike. In essence, they [government] would protect the most vital supplies (i.e. power stations, gas works, iron, steel, railways and house coal) by spreading the shortage over the rest of industry.'[187]

Aubrey Jones, the new Minister of Power, argued that high economic growth meant that increased oil use and nuclear power would not reduce the country's dependence on coal and the miners until 1970 and he sought Cabinet permission to continue with open cast mining even though this would annoy the NUM and the farmers. Absenteeism had not been reduced by existing incentives, there was no possibility of the NUM accepting foreign workers, and full employment was attracting labour away from the pits. The Cabinet's sense of powerlessness found expression in its concern at the over-optimistic and excessive statements about nuclear power 'which could not have more than a marginal effect for a generation to come.'[188] The obvious means of increasing output, despite increasing mechanisation, was to attract more labour into the pits but this required increased pay. A hike in general pay rates was out of the question at a time when government was increasingly concerned about wages and, in any case, the NCB could not afford it. Transferring labour from surplus to deficit coalfields ran up against the miners' traditional refusal to leave friends and family; transferees would require housing, transitional payments, and resettlement allowances. Also, there was no way of keeping transferred labour in the pits at a time of full employment. Finally, any attempt to attract labour by differential coalfield pay rates would be opposed resolutely by the NUM. These factors reinforced the miners' sense of indispensability and 'The miners have put forward a new wage claim as well as miners' charter containing drastic proposals for shorter working hours and more holidays. In view of the very strong bargaining position held by the miners some concession is inevitable.'[189] Coal shortages and the NUM's potential power led ministers to discuss the possibility of a substantial increase in investment, despite a general stringency in public sector investment, in new seams and mechanisation to replace manpower. However, in January 1956 Jones reluctantly accepted a £5m cut in the NCB's investment programme but warned of the 'connection between the level of investment in the coal industry and the confidence of the miners in the long-term future of the industry.'[190]

Once it was clear labour could not be attracted into the pits, ministerial attention focused on the NCB's ability to realise its ambitious mechanisation programme. However, 'It is now clear that the original Plan for Coal under-estimated the difficulties involved; both the time and the money needed to sink new pits and carry out major reorganisations proved substantially greater than had been foreseen. As a result effective investment has fallen behind that contemplated and *it is only in the last two or three years that it can be said to have got under way*'.[191] Thus, 'miscalculation' and 'inauspicious beginnings' meant 'unfortunately' despite spending £128 million on major schemes only schemes to the value of £11 million had been completed by the end of 1955. Despite this Jones concluded 'there is solid ground for believing that as more and more schemes are completed output will improve'. Eden wrote on his copy; 'When will they be'?[192] Eden and the Cabinet had no option other than to sanction continued investment but they insisted these plans be reviewed annually and expressed disquiet about the failure of investment to deliver output; 'greater emphasis would be laid in future on bringing development schemes to fruition in preference to distributing investment over an excessive number of such schemes.'[193]

Tighter economic criteria were also applied to coal prices. Jones pointed out 'sound commercial principles' required the gas and electricity industries to fix prices to cover operating costs and replace capital but this provision had never been applied to the NCB. The NCB should now be brought into line and be made less dependent on the Treasury. This was challenged by the Minister of Transport, Harold Watkinson, who regarded the state industries as economic regulators and they 'should be required to keep their prices as low as possible, in the interests of the general policy of price stability.' Peter Thorneycroft, the Chancellor, responded that wage increases were in excess of price increases and prices would have to rise so 'it was important to ensure that rising prices were immediately linked in the public mind with the wage increases which had been responsible for them.' The Minister of Labour, Ian Macleod, expressed concern that substantial price rises to cover capital investment would stimulate a further round of wage increases. A substantial increase in the price of coal 'would be bound to jeopardise the prospects of securing the collaboration of the trade union movement' in combating the wage-price spiral. David Eccles, the President of the Board of Trade, warned that increased prices would damage exports but he would accept them if 'their inflationary effects were kept under restraint by a strict monetary policy.' The Cabinet agreed to increase prices in order to cover operating costs but not to cover capital investment.[194] The Cabinet approved a 6/6d (37.5p) per ton increase in pit-head prices.

Faced with growing scepticism from the government both the NUM and NCB struggled to find the common ground needed to stabilise their relationship. This attempt to revivify a sectoral elite consensus was complicated by constant pressure from the NUM's members for better pay and conditions and contrary pressure from government on the Board to cut costs. These pressures can be seen in the downgrading of the second Miners' Charter. The original 1946 Charter was a product of its time whilst that of 1955 seemed anachronistic and was purely symbolic as the NUM's leaders knew it would not be implemented by the NCB. Except for the seven and a half hour day, the 1946 Charter had been realised and the reduction in hours was accepted by the NCB as a legitimate objective to be realised over an unspecified period of time.[195] At the 1956 NUM Conference several delegates urged the NEC to take resolute action to secure the Charter. Jim Hammond, for example, argued that 'There comes a time in protracted negotiations when you say that people are playing for time, and therefore we have to take the bull by the horns and help them to think properly.' A second Lancashire delegate commented, 'we have often said and emphasised the power that the Miners' Union wields, both in the political and in the economic sphere. We have an organisation *par excellence*. We have a monopoly in an industry enjoyed by no other trade union in the country.'[196] The NEC recognised the weight of the discontent but 'we should remember that we have our responsibilities...we should endeavour to assure the nation and the industry that we will do our best in this question of seeking control those matters which give rise to stoppages of work.'[197] The resolution was remitted to the NEC.

Notes

[1] Much of this is a myth or exaggerated. On Tonypandy and Churchill's role in the General Strike see P. Addison, *Churchill on the Home Front 1900-1955*, Pimlico 1993, pp. 142-44 and pp. 259-68. Harold Macmillan was reported as saying 'There are three

bodies no sensible man directly challenges: the Roman Catholic Church, the Brigade of Guards and the National Union of Mineworkers.' A. Jay ed., *The Oxford Dictionary of Political Quotations*, Oxford University Press 1966, p. 245. The key word is 'directly'. The quotation is also ascribed to Baldwin.
2. Sir Cuthbert Headlam MP Diaries *D/He 40*, 13 March 1944. Durham County Record Office.
3. N. Harris, *Competition and the Corporate Society. British Conservatism, The State and Industry 1945-1964*, Methuen 1972, p. 85.
4. NUM, *Report of a Special Conference*, 5 April 1951, p. 20.
5. Quoted in Conservative Council of Trade Unions Bulletin 50 (August 1951), p. 1. *CCO 4/4/127* Conservative Party Archives, Bodleian Library, Oxford. Horner enjoyed a reputation amongst Conservatives in the 1940s and 1950s similar to that of Arthur Scargill in the 1970s and 1980s.
6. Socialist Record Exposed, *Notes on Current Politics* 6, 7 April 1947, p. 6.
7. The Coal Situation, *Notes on Current Politics* 17, 29 September 1947, pp. 21-22.
8. Conservative Party Conference, *Notes on Current Politics* 19, 27 October 1947, pp. 2-3.
9. Conservative Party Conference, p. 10.
10. Conservative Policy, *Notes on Current Politics* 15, 1 August 1949, p. 27.
11. Conservative Industrial Policy, *Notes on Current Politics* 9, 19 May 1947, p. 18.
12. Conservative Policy, p. 28.
13. Conservative Conference 1949, *Notes on Current Politics* 21, 7 November 1949, p.7.
14. 5s H.C. Debs 494, 19 November 1951, col. 24. Lloyd repeated this in a written answer (494, 26 November, col. 100) and Gaitskell's formulation can be found in 5s H.C. Debs 485, 29 November 1948, col. 1752.
15. 5s H.C. Debs 513, 23 March 1953, col. 484.
16. 5s H.C. Debs 544, 20 July 1955, col. 473.
17. R. Taylor, *The Trade Union Question in British Politics. Government and Unions since 1945*, Blackwells/ICBH 1993, p. 74. See also, A.J. Taylor, 'The Trade Union and the Party' in A. Seldon and S. Ball eds., *The Conservative Century. The Conservative Party Since 1900*, Oxford University Press 1994, pp. 513-17.
18. 5s H.C. Debs 520, 3 November 1953, col. 23.
19. A. Roberts, *Eminent Churchillians*, Weidenfeld & Nicolson 1994, pp. 211-85 for a critical analysis of Moncktonism and its political legacy.
20. Autobiographical Fragments, p. 6. *Dep. Monckton 49*. 40. Bodleain Library, Oxford.
21. CC (52) 75th Conclusions, 31 July 1952. *CAB 128/24*.
22. C (57) 182, 30 July 1957. The Economic Costs of Industrial Disputes. Note by Officials, para 85, p. 17. *CAB 129/88*.
23. Harold Macmillan to Walter Monckton, 16 June 1952. *Dep. Monckton 2*. fs 230-31.
24. 5s H.C. Debs 544, 20 July 1955, col. 475.
25. C (51), 5, 3 November 1951. Coal. Note by the Secretary of State for the Co-ordination of Transport, Fuel and Power *CAB 129/28*, and CC (51) 4th Conclusions, 5 November 1951. *CAB 128/53*.
26. Lloyd to Eden, 5 December 1955. *PREM 11/1229*.
27. Cherwell to Churchill, 1 August 1952, and Norman Brook to Churchill, 20 December 1954. *PREM 11/1054*. A civil nuclear power programme based on MAGNOX reactors would also provide plutonium for the bomb programme.
28. Lloyd to Churchill, 1 January 1954. *PREM 11/1744*. As a former petroleum minister Lloyd was well aware of the potential of the oil industry and its desire to break into coal's markets.

[29] J.H. Bamburg, *The History of the British Petroleum Company. Vol.2 The Anglo-Iranian Years*, Cambridge University Press 1994, for detail. *POWE 17/81* contains a large amount of material on oil substitution.
[30] Middlemas, *Power, Competition and the State. Vol.1*, pp. 206-07.
[31] Industrial Action for Political Ends, 30 April 1952. *CRD 2/7/11*. This analysis was in response to a CRD officer compiling 13 foolscap pages of newspaper reports advocating the use of strikes for political purposes, complete with quotes from union leaders (including Horner) advocating such action. Industrial Action for Political Ends. Mr Stebbings to Mr Fraser, 3 March 1952. *CRD 2/7/11*.
[32] *TUC Annual Congress Report 1952*, p. 300.
[33] B.J. Evans and A.J. Taylor, *From Salisbury to Major. Continuity and Change in Conservative Politics*, Manchester University Press 1996, pp. 81-96.
[34] A. Montague Browne, *Long Sunset*, Cassell 1995, p. 130. Emphasis added.
[35] *NUM (ACR) 1952*, p. 23.
[36] Meeting with the NCB, *NUM (EC)*, 9 January and *CINC* (45), 15 September 1953.
[37] *The Times*, 20 November 1952.
[38] Ashworth, *History of the British Coalmining Industry*, p. 636.
[39] Sir Gerald Nabarro, *NAB 1. Portrait of a Politician*, Robert Maxwell 1969, p. 209.
[40] CC (53) 19th Meeting, Minute 3, 12 March 1953. *CAB 128/36*.
[41] *5s H.C. Debs 512*, 12 March 1953, col. 149 (Lloyd) and col. 1478 (Monckton).
[42] Motion by Tory MPs, *NUM (EC)*, 12 March 1953.
[43] *NUM (ACR) 1953*, p. 124. Several NUM Areas had supported calls for an internal inquiry into the NCB's structure and operation.
[44] A. Seldon, *Churchill's Indian Summer. The Conservative Government 1951-55*, Hodder and Stoughton 1981, p. 236, and P. Goodhart, *The 1922. The Story of the 1922 Committee*, Macmillan 1973, pp. 165-66.
[45] *5s H.C. Debs 544*, 20 July 1955, col. 406 (Farey-Jones), col. 419-20 (Jones), col. 432 (Lancaster), col. 452 (Braithwaite), and col. 473 (Lambton).
[46] Foreign Office telegram No.85, 21 July 1955. *PREM 11/1744*.
[47] Nabarro, *NAB 1. Portrait of a Politician*, p. 209.
[48] A. Horner, *Incorrigible Rebel*, Macgibbon & Kee 1960, p. 192.
[49] CC (51) 8th Conclusions, 19 November 1951. *CAB 129/48*.
[50] Addison, *Churchill on the Home Front 1900-1955*, pp. 413-14.
[51] *NUM (ACR) 1951*, p. 23 and pp. 126-27.
[52] *NUM (ACR) 1950*, p. 106.
[53] *NUM (ACR) 1951*, pp. 196-99 for the full debate.
[54] NUM, *Report of a Special Conference Report*, 18 December 1952, p. 14. Speech by Horner.
[55] Meeting with the National Coal Board, *NUM (EC)*, 9 January 1953.
[56] Appendix VII. Minutes of Special Sub-Committee on the Control of the Mining Industry and National Coal Board Administration, 3 May 1951, *NUM (EC)*, 4 May 1951. Emphasis added.
[57] NUM, *Report of a Special Conference*, 14 March 1952, p. 9.
[58] *Report of a Special Conference*, p. 15.
[59] *Report of a Special Conference*, p. 21.
[60] Middlemas, *Power, Competition and the State, Vol. 1*, p. 228.
[61] Lord Swinton to David Gammans, 21 August 1952. *PREM 11/314*.
[62] *NUM (ACR) 1952*, p. 23.

63 CC (53) 2nd Conclusions, 14 January 1953. *CAB 128/26*. Brook to Churchill, 4 February 1953. *PREM 11/1746*.
64 CC (53) 7th Conclusions, 5 February 1953. *CAB128/26*.
65 C (53) 49, 6 February 1953. Coal Prices. Memorandum by the Minister of Labour. *CAB129/58*.
66 Coal Prices. Lord Cherwell to Churchill, 6 February 1953. *PREM 11/1746*.
67 CC (53) 8th Conclusions, 10 February 1953. *CAB128/26*.
68 CC (53) 11th Conclusions, 12 February 1953. *CAB128/26*.
69 Diary 101, 16 March 1953. *MS Woolton 3*. Woolton Papers. Bodleian Library, Oxford.
70 *NUM (ACR) 1952*, p. 49. Emphasis added.
71 *NUM (ACR) 1952*, pp. 72-6.
72 *NUM (ACR) 1952*, pp. 111-127. Emphasis added. In the vote the Derbyshire and Northumberland delegations were split but had to cast all their votes one way. Resolution 35 was lost by 259,000 to 409,000 which indicates a considerable residuum of support for a more left-wing policy.
73 *NUM (ACR) 1952*, p. 131.
74 CC (52) 75th Conclusions, 31 July 1952. *CAB 128/55*. The TUC protested to Churchill who overruled Monckton.
75 *JNNC (73)*, 22 July, *NUM (EC)*, 6 August, and *JNNC (76)*, 30 October 1952. There were a spate of unofficial strikes in the Scottish and Lancashire coalfields during this period.
76 *The Times*, 31 October 1952.
77 NUM, *Report of a Special Conference*, 7 November 1952, p. 5, p. 12, and p. 21. Thirty miners from Scotland picketed the conference and jeered Lawther.
78 A daywageman was a mineworker who was paid a set daily rate for his job unlike pieceworkers. At this time daywagemen were largely employed away from the coalface and compared to pieceworkers (essentially engaged in coal production and face development) were relatively poorly paid. The political consequences of wage restructuring will be discussed in Volume 2, chapter 1.
79 *The Times*, 14 November 1952.
80 *NUM (EC)*, 13 November, and *JNNC (78)*, 13 November 1952.
81 *JNNC (79)*, 14 November 1952.
82 *The Times*, 18 December 1952.
83 NUM, *Report of a Special Conference*, 18 December 1952, pp. 18-19.
84 *Report of a Special Conference*, p. 21 and p. 23.
85 *Report of a Special Conference*, pp. 31-32.
86 Situation in the Industry, *NUM (EC)*, 8 January 1953.
87 Meeting with the National Coal Board, *NUM (EC)*, 9 January 1953.
88 CC (53) 3rd Conclusions, 20 January 1953. *CAB 128/26*.
89 Meeting with the National Coal Board, *NUM (EC)*, 22 January 1953.
90 CC (53) 4th Conclusions, 22 January 1953. *CAB 123/26*.
91 NUM, *Report of a Special Conference*, 27 January 1953, p. 13 and p. 16.
92 *Report of a Special Conference*, p. 19.
93 *Report of a Special Conference*, p. 24.
94 *Report of a Special Conference*, pp. 25-26.
95 *Report of a Special Conference*, p. 26 and p. 27.
96 *Report of a Special Conference*, pp. 31-3.
97 *Report of a Special Conference*, p. 38.
98 Wages -- Meeting with the NCB, *NUM (EC)*, 12 February 1953.

99 CC (53) 8th Conclusions, 10 February 1953. *CAB128/36*.
100 CC (54) 23rd Conclusions, 31 March 1954. *CAB 128/37*.
101 C. (53) 193, 7 July 1953. Coal Supply Position. Memorandum by the Secretary of State for Co-ordination of Transport, Fuel and Power and the Minister of Fuel and Power. *CAB129/61* and C.C. (53) 41st Conclusions, 9 July 1953. *CAB 128/26*.
102 C. (53) 199, 20 July 1953. Coal Prospects. Memorandum by the Secretary of State for Co-ordination of Transport, Fuel and Power and the Minister of Fuel and Power. *CAB128/27*.
103 The Coal Committee (GEN 445) was set up by the Cabinet on 6 October 1953 to submit proposals for a comprehensive policy on coal. For composition and terms of reference see, Elements of a Coal Policy. Memorandum by the Chancellor of the Exchequer. GEN 445 2, 12 October 1953, and Terms of Reference and Composition. GEN 445 1st Meeting, 9 October 1953. *CAB 130/95*.
104 Elements of a Coal Policy. GEN 445 1st Meeting, 14 October 1953. *CAB 130/95*.
105 Fuel Policy. Memorandum by the Minister of Fuel and Power. GEN 445/4, 3 April 1954, para 4. *CAB 130/95*.
106 Fuel Policy, para 16. *CAB 130/95*.
107 Fuel Policy, para 24. *CAB 130/95*.
108 GEN 445/3rd Meeting. Minute 1, 9 May 1954, p. 1. *CAB 130/95*.
109 GEN 445/3rd Meeting, p. 3. *CAB 130/95*.
110 C.(54). 163, 18 May 1954. Fuel Policy. Memorandum by the Minister of Fuel and Power. *CAB 129/65*.
111 C. (54) 170 20 May 1954. Fuel Policy. Memorandum by the Chancellor of the Exchequer. *CAB 129/68*.
112 Geoffrey Lloyd to Churchill, 24 May, and Norman Brook to Churchill, 26 May 1954. *PREM 11/1744*.
113 Norman Brook to Churchill, 1 June 1954. *PREM 11/1744*.
114 CC (54) 37th Conclusions, 2 June 1954. *CAB 128/27*.
115 E.A. (54) 86, 26 July 1954. Fuel Policy. Note by the Chancellor of the Duchy of Lancaster and Minister of Materials. *PREM11/1744*.
116 Colville to Churchill, 27 July 1954. *PREM11/1744*.
117 Lloyd to Churchill, 7 December 1954. *PREM11/1744*, and C.C. (54) 84th Conclusions, 9 December 1954. *CAB128/27*
118 E.A. (55) 60, 4 April 1955. Fuel Policy. Memorandum by the Minister of Fuel and Power. *PREM11/1744*. CC (55) 7th Conclusions, 27 January 1955. *CAB128/28*.
119 C (55) 22nd Conclusion, 9 March 1955. *CAB 128/28*.
120 C (55) 24th Conclusions, 19 July 1955. *CAB 128/29*.
121 5s *H.C. Debs 544*, 20 July 1955, col. 387.
122 5s *H.C. Debs 544*, 20 July 1955, col. 466.
123 CM (55) 17th Conclusions, 23 June 1955. *CAB 128/29*.
124 Lloyd to Eden, 3 October 1955. *PREM 11/1744*. Eden thought Lloyd was overly pessimistic.
125 Seldon, *Churchill's Indian Summer*, p. 193.
126 Organisational decentralisation did not breach the government's commitment that it would not decentralize collective bargaining or promote inter-coalfield competition. It might, however, be perceived by the NUM as the thin end of a wedge which might culminate in denationalisation. 5s *H.C. Debs 494*, 19 November 1951, col. 24 for the government's position.
127 *NUM (EC)*, 12 March 1953.

[128] Minutes of a Special Sub-Committee on the Control of the Mining Industry and National Coal Board, 17 June 1953.
[129] *NUM (ACR) 1953*, p. 21.
[130] *NUM (ACR) 1953*, pp. 122-23.
[131] *NUM (ACR) 1954*, pp. 122-23.
[132] For example, EA (54) 86, 26 July 1954. Fuel Policy. Note by the Chancellor of the Duchy of Lancaster and Minister of Materials, para 6. *CAB129/65*.
[133] Bracken to Beaverbrook, 7 January 1953. R. Cockett ed., *My Dear Max. The Letters of Brendan Bracken to Lord Beaverbrook 1925-1958*, The Historian's Press 1990, pp. 133-34.
[134] C. (53) 199. Coal Prospects. Memorandum by the Secretary for the Co-ordination of Transport Fuel and Power and the Minister of Fuel and Power. 20 July 1953, p. 3. *CAB 129/61*.
[135] C. (53) 273. Coal. Memorandum by the Minister of Fuel and Power. 2 October 1953, p. 2. *CAB 129/61*.
[136] CC (53) 55th Conclusions, Minute 4. 6 October 1953. *CAB 128/36*. Emphasis added.
[137] GEN. 445/1, 14 October 1954. Elements of a Coal Policy, p. 1. *CAB130/95*.
[138] GEN 445/4, 3 April 1954. Fuel Policy. Memorandum by the Minister of Fuel and Power, para 33. *CAB130/95*. Briquetting was a process similar to making coke whereby small pieces of coal, or even coal dust, was combined into larger lumps, or briquettes. There was considerable consumer resistance to briquettes.
[139] C. (54) 163, 18 May 1954. Fuel Policy. Memorandum by the Minister of Fuel and Power, para 4. *CAB 129/65*. Emphasis added.
[140] C. (54) 170, 20 May 1954. Fuel Policy. Memorandum by the Chancellor of the Exchequer, para 4. *CAB 129/65*.
[141] Norman Brook to Churchill, 26 May 1954, *PREM 11/1744*. Fuel Policy C. (54) 163, 170, and 171, para 5. *CAB129/65*.
[142] Ashworth, *The History of the British Coal Industry*, pp. 191-92.
[143] *Report of the Advisory Committee on Organisation*, National Coal Board 1955. Hereafter, *Fleck Report*.
[144] Ashworth, *The History of the British Coal Industry*, p. 194.
[145] *Fleck Report*, para 5, p. 2.
[146] *Fleck Report*, para 27, p. 7.
[147] Appendix IX 'Review of the National Coal Board's Organisation', para 6, *NUM (EC)*, 14 October 1954. The NEC also pointed out that whereas originally there had been two 'labour' representatives on the Board, there was now only one.
[148] 'Review of the National Coal Board's Organisation', para 7.
[149] 'Review of the National Coal Board's Organisation', para 12.
[150] *Fleck Report*, para 121, pp. 25-26.
[151] Fuel Policy. Memorandum by the Minister of Fuel and Power, EA (55) 60, 4 April 1955. para 2. *PREM11/1744*.
[152] EA (55) 60, 4 April 1955, para 20 (I). *PREM 11/1744*. See also CM (55) 24th Conclusions Minute 1, 19 July 1955. *CAB 129/29*.
[153] Ashworth, *The History of the British Coal Industry*, p. 196.
[154] Report on the Advisory Committee on the Organisation of the National Coal Board, *NUM (EC)*, 22 March 1956.
[155] *NUM (ACR) 1956*, pp. 120-22.
[156] *NUM (ACR) 1956*, p. 125.
[157] *NUM (ACR) 1956*, pp. 138-39.

158 Coal Board Appointments, *NUM (EC)*, 13 December 1956.
159 Industrial Relations Member, *NUM (EC)*, 14 February 1957.
160 Industrial Relations Member, *NUM (EC)*, 28 February 1957. My emphasis.
161 Industrial Relations Member, *NUM (EC)*, 31 January 1957.
162 *NUM (ACR) 1957*, p. 146 (Watson), p. 147 (Pratt and McKendrick), p. 149 (Wynn).
163 *NUM (ACR) 1957*, pp. 150-51.
164 *NUM (ACR), 1953*, p 67.
165 *NUM (ACR), 1953*, p. 77.
166 *NUM (ACR), 1953*, p. 79.
167 *NUM (ACR), 1953*, pp. 90-91.
168 JNNC (88), 13 October, and Wages, *NUM (EC)*, 29 October 1953.
169 Wages-Meeting with the NCB, *NUM (EC)*, 7 January 1954. The offer was to extend the EHA to 30 April 1955 and increase underground minima to £7.75 and the surface minima to £6.75. The only option to recommending acceptance was a reference to the NRT. The Special Conference (27 January 1954) voted by 537,00 to 227,000 to accept the NCB's offer which was accepted by the Areas by 542 in favour, 221 against.
170 Miners' Wage Claim. Lloyd to Churchill, 1 January 1954. *PREM 11/1744*. Houldsworth believed he could secure a public declaration from the NUM agreeing to link pay to a productivity increase of 2.5 per cent in 1954. The election was the result of Lawther's retirement.
171 *NUM (ACR) 1955*, p. 41 (J.R.A. Machen, Yorkshire).
172 Unsigned note to Woolton, 21 September, and 189 (reply to 188) 24 October 1955. *MS Woolton 22*.
173 Yorkshire Coalfield Dispute 1955. *PREM 11/1030*. The existence of this file demonstrates eloquently the political sensitivity of coal and its abiding interest to Prime Ministers who were acutely conscious of their dependence on coal.
174 *NUM (ACR) 1955*, pp. 80-81.
175 *NUM (ACR) 1955*, p. 57.
176 H. Macmillan, *Memoirs. Riding the Storm, 1956-1959*, Macmillan 1971, p. 55.
177 *NUM (ACR) 1955*, p. 97.
178 *NUM (ACR) 1955*, p. 203.
179 Meeting with the NCB, *NUM (EC)*, 10 November 1956.
180 CP(56) 21, 26 January 1956. Fuel and Power Prospects. Memorandum by the Minister of Fuel and Power, para 15, p. 4. *CAB 129/69*.
181 Aubrey Jones to Iain Macleod, 14 February 1956. *PREM 11/1744*. Aubrey Jones was Minister of Fuel and Power (December 1955 to January 1957), Macleod was Minister of Labour and National Service.
182 JNNC (107), 14 December 1955, and Meeting with the NCB, *NUM (EC)*, 16 February 1956. Negotiations were slowed by the death of Sir Hubert Houldsworth and James Bowman took over negotiations at short notice.
183 A.A. Jarratt to Eden, 16 February 1956. *PREM 11/1744*.
184 CINC (61), 12 April 1956, and *NUM (ACR) 1956*, p. 23.
185 Middlemas, *Power, Competition and the State. Vol. 1*, p. 288.
186 Record of a Conversation, 19 August 1955. *PREM 11/1744*.
187 Noel Martin to David Pitblado, 31 October 1955. *PREM 11/1744*.
188 CM (55) 8th Conclusions Minute 9, 31 January 1956. *CAB128/30*.
189 CP (56) 21, 26 January 1956. Fuel and Power Prospects. Memorandum by the Minister of Fuel and Power, para 15. *CAB129/79*. This assessment was prepared at Eden's request.

[190] CM (56) 6th Conclusions, 24 January 1956. *CAB 128/30*.
[191] CP (56) 87, 24 March 1956. Coal Investment. Memorandum by the Minister of Fuel and Power. para 5. *CAB129/80*. Emphasis added.
[192] Coal Investment. *PREM 11/1230*.
[193] CC (56) 25th Conclusions, 27 March and 32nd Conclusions, 3 May 1956. *CAB128/30*.
[194] CC (57) 38th Conclusions, 6 May 1957. *CAB128/30*.
[195] *NUM(ACR) 1956*, p. 64.
[196] *NUM (ACR) 1956*, p. 68.
[197] *NUM (ACR) 1956*, p. 73.

Chapter 5

The Politics of Industrial Decline

Introduction

This chapter examines the effect on the NUM of the fall in demand for coal in 1957 as the energy market changed and the coal substitution policies put in place between 1951 and 1955 began to work. Initially this decline was thought to be temporary, the result of a cyclical recession, but it soon became clear the changes were structural and for the first time since 1947 pits closed on economic grounds. This provoked a debate in the NUM about how it should respond to this new situation: lobbying and pressure group politics or industrial action? The latter was ruled out of the question and as the Macmillan government rejected the NUM's policy, the only hope was the election of a Labour government. The NUM believed that Labour government of 1964 was committed to creating a 200m ton industry. They were to be disappointed.

Markets and Strategy

The Conservative government's move to incomes policy generated some hostility in the NUM. A resolution from South Wales and Kent to the 1956 Conference represented, according to Bill Paynter, 'a declaration by the Union of the form of action we intend to take in order to counter the policy of the present Tory Government. It lines up the miners with other sections of industrial workers in this country into a mighty front of opposition to defeat the Tory Government.' Industrial action to improve wages 'can make possible the acceleration of the day when this Government will be defeated.' Paynter conceded that by accepting the resolution the NUM might be endangering public ownership, but public ownership should be 'a vehicle that would be capable...of stimulating economic and social progress [not] the handmaiden to privately owned industry, to enhance the profits of such privately owned industry.' The NEC accepted the composite's spirit but Jones warned 'that if this resolution means the use of the industrial machine for the purchase of achieving political ends [it is] not the sort of motion that this Conference should support.'[1]

A call that the NUM re-evaluate compulsory arbitration and its participation in the NRT was potentially much more significant than the South Wales/Kent resolution. Whilst it was unlikely that a majority of miners would support a national strike, although as this proposition was not tested one cannot say this with certainty, many were willing to contemplate withdrawal from the NRT. The problem was that the NRT was one of the aspects of the post-nationalisation industrial relations system which was supposed to distinguish it from the anarchy of private ownership. The NUM had 100 per cent membership and coal remained central to the economy and, as a result, 'We have had in recent years more power than ever at any time before in our history' and so the NUM could have achieved more without the NRT. The NEC supported the resolution because it called for a review of the twenty six awards made since 1944. The NEC remained convinced of the NRT's utility and to question it was to question coal's entire industrial relations system as the NRT was symbolic of both the industry's changed nature and the NUM's wider responsibilities. Thus, 'We have always known we could get everything we demanded...but this organisation has demonstrated...the highest sense of social responsibility of any other organisation in the country...we would wreck this country. Of course, we, like Samson, could tear the pillars down, but we would be in the wreckage.'[2]

Unrest in the industry contributed to calls that the NUM end voluntary Saturday working. Reflecting the post-Fleck managerial climate the NCB stood firm insisting on its continuation; the Prime Minister minuted, 'I am glad the board is standing firm. We must do all we can to help them'.[3] Government was in something of a quandary concerning wages. Macmillan, Eden's Chancellor, knew from his contacts in the unions and TUC that any wage freeze 'would be deeply resented and strenuously resisted' but many union leaders recognised the government's difficulties and were anxious to help as long as they were not open to the charge of being 'Tory stooges'.[4] During the long frustrating negotiation of the 1956 NUM wage claim industrial action was not even considered despite the NCB's intransigence.[5] This resulted in an unprecedented event which reflected the growing instability of post-war coal industry politics. The NUM's negotiators secured some concessions from the NCB but not enough to satisfy the membership who, for the first time since nationalisation, rejected the draft agreement by 22 to 451 in an Area vote. The NEC was also under pressure from Nottingham, South Wales,

Scotland, and Northumberland to seek an increase of 3s 3d on all national grate rates as a result of a sharp increase in the cost of living and the fuel shortages caused by the Suez Crisis.[6] By 1957 the NEC was conscious that the industry seemed to becoming less responsive to their needs, a change which was symbolised by the appointment of a non-miner as NCB Labour Director.[7] Meeting to discuss the situation the NEC decided to make no recommendation to the Special Conference called to discuss the situation in the industry, a decision which reflected the NEC's growing strategic bankruptcy.[8]

A South Wales delegate told the Special Conference 'there is too much shilly-shallying' with the NCB and government and as a result he believed the men in the pits were losing confidence in their leaders so 'it is now time we took the gloves off. They feel it is time to get tough...You have got the power, and you are being asked to use it.'[9] Sam Bullough (Yorkshire), who was no militant, believed 'we must take serious note of the feelings of the rank and file in this industry.'[10] These sentiments, expressed by both 'moderate' and 'militant' NUM Areas, were deeply worrying as they seemed to represent an emerging general challenge to the culture and ethos of public ownership. If these sentiments ran out of control the NEC foresaw the collapse of both nationalisation and the NUM. Horner said:

> There is one thing however which I do not think is fully appreciated in this Conference. We are not living in the days of the old coal owners. We are dealing with an organisation which we helped to create, the nationalised coal industry, and judged purely from the angle of benefits to the men...life with a nationalised coal mining industry is enormously better than life could ever be under the old owners.[11]

This strategic debate continued at a second Special Conference called ostensibly to discuss the NCB's revised offer. Horner again deprecated the speed with which some mineworkers had resorted to unofficial strike action arguing the NUM's central object must be *'to try to assist the [NCB] to justify its existence...*the [NCB] would be the first to confess it, that they depend absolutely on the goodwill of the Union'. In response to a question from Glynn Williams, Horner agreed that under existing agreements the only option open to the NUM other than accepting the NCB's offer was reference to the NRT whose decision the NUM was obliged to accept. A reference to the NRT would be unpopular but the issue at stake 'is not big enough to have a national crisis about. There are plenty of bigger things than this, and therefore *it is not worth the destruction of the machinery*'.[12] The delegates disagreed and on a card vote voted by 370,000 to 345,000 to reject the NCB's offer. Jones, the NUM President, interpreted this (to the fury of the delegates) as a *de facto* instruction to refer the dispute to the NRT because the conciliation agreement permitted no other course of action, but Sam Watson proposed the Areas vote on the offer. This was agreed.

The Areas voted by 266 to 466 to reject largely because according to the NEC, the NCB had linked the abolition of the Bonus Disqualification to the employment of Hungarian refugees.[13] Attitudes were worsened by the NCB conceding a 5 per cent increase to the Deputies and Overmen which, the NEC noted, was made after 'unconstitutional action and threats' by NACODS members. Refusing to accept the NCB's offer as final 'the negotiators were faced with a breakdown on the wages question. However, *what was more dangerous* was that they were faced with a breakdown in the relationship between the Board and the Union.' The NUM were convinced the NCB was doing the government's dirty work and although the Board denied this, it did concede there were 'wider national interests which it was the Board's responsibility to consider.'[14] The final offer was of a general wage increase of 5 per cent and the abolition of the Disqualification Bonus except in the case of unofficial strikes and stoppages, and the NEC recommended acceptance of this to a Special Conference. Horner was adamant that the NCB's intransigence was the product of government interference because 'the [NCB] is a Governmental organisation, and for the [NCB] to step out of line with the general tendencies in industry was something very serious for the [NCB].' The offer was unsatisfactory but it was the best that could be achieved given the limitations imposed by NCB policy and the NRT. This assessment was accepted by the delegates although Mick Kane (Derbyshire) warned that 'sooner or later we have got to challenge the Tribunal...We can say the same as the shipbuilders have said, that it is not necessary to accept the findings of the Tribunal', and Dick Main (Northumberland) admitted that:

> We have always had secret misgivings in regard to the [NRT]...we have to make more demands to improve wages...Surely the time of our greatest economic strength is the time to tell people that we do not want just an existence wage, but that we want a living wage.[15]

Delegates voted by 472,000 to 262,000 and the Areas by 495 to 237 to accept the NCB's offer.

These controversies resurfaced at the 1957 NUM Conference. Resolution 11 (Derbyshire) was a direct result of the politicisation of collective bargaining by government incomes policy. As government policy was pushing up the cost of living so the NUM should respond 'and it is an open secret known to the men in the coalfield that if the [NUM] pursues a claim with sufficient determination, this Government cannot stand against us.' Horner agreed with most of the resolution's sentiments but warned:

> We cannot fight Parliament as a Union, but we can fight to cancel out the effect of what a reactionary Tory Government does, and we know our power... We exercise it with great restraint, because we know that the whole industrial apparatus of this country depends on our people.[16]

Horner was adamant that the NUM had to act 'with a full understanding of the effect of our actions upon the British people. We [were] isolated for many years from the general community of the British people.' Nationalisation signified the miners' full integration into society and by 1957 the fruits of public ownership were becoming apparent, 'under any test – production, reorganisation, wages of the men, conditions of the men, or welfare facilities – nationalisation has provided something which is far superior, in spite of criticism to anything we could possibly have got under private ownership.'[17] These convictions were about undergo a severe test.

In 1957 demand for coal fell sharply. The NUM blamed this on the macro-economic and energy policies of the Conservative government which were closing pits and blocking improvements in the NUM's members' pay and conditions, so forcing the NCB into confrontation with the NUM. Government had blocked the exports of coal and profitable markets had now been lost to US coal, it had refused to limit oil imports, the Clean Air Act (1957) and the railway's conversion to diesel haulage were damaging the coal industry. The NUM called for a joint approach to the government with the NCB in an attempt to divert their members' frustration away from the Board by portraying the NUM and NCB as equal victims of government policy. On the other hand, the NEC did not want their members' frustration to be directed at government. The NUM's solution was 'a planned economy, not an attack on full employment' but this required a Labour government so until there was a Labour government the NUM 'would resist any worsening of its members conditions' by co-operating with the NCB in the wider interests of the coal industry.[18] Throughout 1958 the NUM pursued orthodox interest group politics using the Miners' Parliamentary Group and its relationship with the NCB to seek to influence the Ministry of Power's policy. This produced no change in the industry's position as the Minister rejected the NUM's call for an end to opencast mining and cuts in oil imports because 'regard must be paid to consumer choice, and that if customers preferred oil to coal, this should be made available to them.'[19] The failure of pressure group politics led to calls for industrial action which were criticised vehemently by NUM leaders. The view that 'industrial workers and trade unions should use their unquestioned industrial and economic power in order to bring down the government' was unconstitutional and dangerous and workers 'must place full reliance upon political democracy.' Failure to do so 'would reduce the possibility of an outstanding Labour Party success at the next and subsequent General Election.'[20] The only legitimate solution to the industry's problems was electoral, the election of a Labour government.

The industry's crisis had obvious implications for collective bargaining. The NCB had already refused to consider any claim for increased wages and the NRT had rejected a claim for a 40 hour week for surface workers and a flat rate increase of 10 shillings (50p) in all wage rates. The NUM was obliged to refer any disputed claim to the NRT and its response in the event of a rejection by the NRT was debated at the 1958 Conference. The debate was notable for a remarkable speech by Horner which expressed eloquently the NEC's post-nationalisation dilemma. The NUM would seek a 15s (75p) a week increase for daywagemen 'intending to use every justifiable means at our command' and he warned 'we have exercised a restraint and a degree of responsibility in this industry through the years when we were completely in command of the situation.' Coal's problems had led some Conservatives to express 'the view that this is the time to teach the miners a lesson. All right, if they feel like a repetition of 1926 they can have it – if that is their desire.' It was not, of course, and Horner admitted there was no possibility of a repeat of 1926 when he declared the NUM's purpose was 'to secure the maximum benefits for our own members with the least possible hurt in getting them...if any [Conservatives] think, because there is a superfluity of coal at the moment, that the time has come to throw down the gauntlet to this organisation, I am sure you would be ready for the challenge.'[21] Will Paynter explained South Wales was withdrawing its wage resolution to maximise unity and would support the NEC's call for a substantial increase of unspecified amount. Paynter argued the NCB was party to a government inspired attempt to hold down wages which the NUM (and the rest of the union movement) had to resist. However, he drew a distinction between conflict over wages which could be resolved by collective bargaining and an attack on public ownership which 'would produce a situation, maybe, where we would be justified in using industrial action to defeat the political aim of that kind.'[22]

As the government had no intention of attacking public ownership, there was no need for industrial action.

A more serious threat was yet another a resolution (from Yorkshire) calling for the termination of the compulsory arbitration clause in the Conciliation Scheme. The objective of this was to increase the NEC's bargaining power by giving it the option to consult the membership on the union's response to a disputed claim. This was important as the resolution contended that the Conciliation Scheme was no longer politically neutral but change did not mean a return to industrial action but the reintroduction of 'a weapon we lost ten years ago.'[23] The resolution's seconder (from South Wales) saw change as the logical if delayed response to the election of a Conservative government. From 1951 onwards 'there has been a hardening attitude of the Board towards the mining industry...They talk now about closing uneconomic pits...the balance of class forces in this latter period has changed against the miners.' The conciliation system emerged out of the needs of war and nationalisation but the world had moved on and the NRT composed of 'judges...or university dons' could not be neutral because 'If you sat on a tribunal, knowing that traditionally tribunals reflect national policy and knowing that a union is obliged to accept the results of that tribunal, you are not likely to give an increase, knowing that no action can follow? Of course you are not.'[24] Alex Moffat reminded Conference that in 1951 Scotland had submitted a similar resolution only to be told that change was not needed as the NCB was different. Not only was the NCB clearly an agent of government but the conciliation system was not conciliating but was breeding frustration. If there were further rejections by the NRT 'it will not matter whether we call for official strikes or not. The miners will come out themselves.'[25] Kent NUM argued that the NUM's gains since 1947 were the result of the high demand for coal but the market was changing reducing the NCB's incentive to negotiate. Despite the decline in coal demand the NUM still had great potential power so 'let us show what we have in our fist when we meet the Board, and then we believe the Board will treat this Union seriously.'[26]

Faced by certain defeat the NEC asked for the resolution to be remitted because it was likely that the Conservative government would be defeated in the next election and a Labour government would be more sympathetic to the NUM. Furthermore, if the NEC felt withdrawal from the NRT was warranted then it would recommend withdrawal at the appropriate time and whilst Horner conceded the membership were frustrated but 'we cannot manage the affairs of this great Union by waves of sentiment or frustration.'[27] The resolution could destabilise the entire conciliation system and Jones implied the resolution would be driven through by the votes of the biggest Areas against the wishes of the NEC which represented the entire union.[28] Kent's amendment was lost, Yorkshire's resolution was approved by 83 to 58. By the Autumn of 1958 the traditional official line against using industrial action for political purposes had been reasserted. Horner warned delegates to a Special Conference that 'As long as we have a Parliamentary system and the unions are subservient to Parliament, and unless we are prepared to challenge Parliament, then we have to try to change the policy of the country by changing the Government of the country.'[29]

The 1958 wage agreement was, as the NEC feared, rejected by the Areas by 322 to 418. The NUM's options were limited to referral to the NRT which was likely to reject the claim; it could table a new claim but this would mean (in effect) no increase for daywagemen; whilst a ballot on strike action 'would be tantamount to an attack on Nationalisation *and did not warrant serious consideration.*' Opinion favoured the NRT option.[30] The reasoning behind this was that the NRT's members would perceive the disturbed state of the industry and, with the tacit consent of ministers, would make some concession in the national interest to assuage the unrest. Paynter, Horner's successor as General Secretary, tried to re-orient the union's policy by arguing the NUM was affected by a wider crisis of capitalism reflected in the Macmillan government's attempt to regulate wages. The Council on Prices, Productivity and Incomes (CPPI), known as the Cohen Council after its chair, Lord Cohen (a High Court judge), was the government's response to a recommendation of the Courts of Inquiry into the engineering and shipbuilding disputes of 1954 and 1957 for an impartial body to place wage claims in their wider inflationary context. CPPI was a failure. Its main consequence was to antagonise the unions who regarded it and the 1956 White Paper, *The Economic Implications of Full Employment* (Cmnd.9725) as the opening moves in a government attempt to intervene in free collective bargaining. Ministers and civil servants were concerned that full employment had resulted in a higher than historically acceptable rate of inflation and were convinced that the unions were the main cause of domestic inflation. Full employment remained sacrosanct but this did not prevent ministers from searching for alternative methods to influence wages, such as a credit squeeze or Ian Macleod's defeat of the London busmen in June 1958. For Paynter the CPPI was an attempt to unite government and employer against union wage claims and he advocated the unions respond in the same way. By the Autumn of 1958

Macmillan's government had settled on a policy: 'In nationalised industries we would require the boards to make savings to balance any wage-increases...They must not increase wages of their own volition.'[31] This had major consequences for the NCB in that it increased further the pressure on management to reduce costs which had implications for pit closures. That government would seek to control public wage sector costs was obvious but this raised the possibility of any wage restraint in mining being interpreted by sections of the NUM as a Conservative challenge to public ownership, so 'if there is any attempt to destroy or weaken nationalisation, maybe, we would be justified in using industrial action to defeat the political aim of that kind.'[32] What is interesting about this statement is its equivocation, *if* government attempted to 'destroy or weaken' nationalisation the miners *might* be justified in using industrial action. As the government had absolutely no intention of attacking nationalisation the threat was meaningless.

CPPI, the belief that the NCB was the government's stooge, and a hardening of NCB attitudes was sometimes portrayed in the NUM as undermining the contract which supported public ownership. Circumstances had changed since 1947 so 'It is now a question of self-defence.'[33] This shift was crystallised by press leaks in December 1958 that the NCB proposed to close 36 pits on economic grounds and several NEC members insisted that this was a government decision despite the Board's insistence that closures were solely their preserve. Closing 36 pits on economic grounds could be plausibly portrayed as an overt attack on both full employment and public ownership in the sense outlined by Paynter at the 1958 Conference. However, a majority of the NEC believed that politicising closures by challenging the government and NCB was not in the industry's or union's long-term interests as the result 'would be tantamount to giving [government] authority over...the conduct of the affairs of the industry.' Moreover, substantial sections of the NUM were willing to see these closures as part of the price to be paid for modernising the coal industry and investment in new, high productivity mines. Therefore the NUM co-operated with the NCB to keep the government out of the industry whilst relying in the longer run on the election of a Labour government.[34]

The nagging problem remained of what to do if no political or electoral solution to the industry's problems emerged. In such a situation Sammy Taylor (Yorkshire) concluded 'modesty and flexibility must go by the board'; Alex Moffat insisted 'that if we are to break the [sic] Tory rule' the

NUM's leaders should show 'the miners that they are ready to fight the policies that have been strangling the economy of this country'.[35]

On the Eve of Destruction

Hitherto closures resulted from reconstruction and poor geology but after 1957 the rationale for closures changed:

> While absolute priority was being given to the quantity of coal production, pits were not closed down simply because they were uneconomic; production was more important than price, and pits were only closed when their resources were virtually exhausted. But as the demand for coal drops, the cost of producing becomes of more importance, and there is a greater incentive to close the pits which are producing the most expensive coal.[36]

In February 1958 the NUM recognised 'the Board were being forced away from the position of expansion' which had underpinned its activities since 1946.[37] Between 1956 and 1958 coal demand fell by 18m tons. This was interpreted as the result of the deflationary measures taken in response to the post-Suez economic crisis but the decline was not cyclical, but structural as the consumption of oil and electricity continued to increase. Percy Mills, the Minister of Fuel and Power, told the Cabinet that 'despite the Board's efforts to reduce their costs by restricting recruitment, ending general Saturday working and closing uneconomic pits, the reduction in consumption is having adverse effects on the financial position of the Board.'[38]

In 1952 the Ridley Committee argued fuel policy should meet four objectives: satisfy in full the community's demand for different fuel and power services at prices which corresponded closely to the costs of production; export fuel with the most gain for the country; promote maximum fuel efficiency; and encourage a pattern of fuel use which gave the best return on resources consumed. Realising these objectives would not be easy because of the 'intricate and variable' nature of the fuel sector which could change dramatically over time.[39] Ridley had dismissed:

> Any attempt to determine by central authority the pattern of the consumers' fuel use [this] would meet formidable problems of enforcement, both because of administrative difficulties and on the grounds of equity...any such intervention would, in our view, unduly restrict the competition...which in the past provided an effective stimulus to enterprise and efficiency.

Ridley concluded 'the right policy, and indeed the only practicable one, is to leave the pattern of fuel use to be determined by the consumer's own choice between competing services.'[40] This became the basis of energy policy making.

In the 1959 debate on the Coal Industry Bill the Minister of Power, Richard Wood, posing the question what was meant by a national fuel policy?, commented 'it has generally been taken to mean a demand for Government intervention...in favour of one fuel and against another.' Wood saw the government's duty as ensuring that the country had varied and secure sources of fuel but would not impose a fuel use pattern as 'It is quite impossible to lay down a simple guide about what fuel is the most economical for general use...the denial of freedom of choice to consumers and the compulsion to use certain fuels rather than others is not only industrially undesirable, but it is humanly immeasurable and indefinable.' Wood pointed out (as had Ridley) that he was constrained by s1(1) of the Ministry of Fuel and Power Act (1945) to promote the efficient and economic use of fuel, so changes in the pattern of fuel use meant that well intentioned decisions could have serious unforeseen consequences. This did not however, equate to a free-for-all.[41] Ministers agreed to increase subsidies for coal stocking, re-examine power station oil contracts, and explore the feasibility of phasing out opencast mining by 1965.[42] Ever ready to grasp at straws the NEC interpreted this as a major breakthrough and called for further government help to facilitate the industry's adjustment.[43] The Minister 'was fully aware of the possible results of the contraction in the coal industry and...was anxious to avoid hardship to individuals and communities' but 'he could not encourage any hope of measures to restrict the use of oil or to force fuel consuming industries to use coal. British industry had to be competitive and, in any case, it was impossible to direct consumers as to the kind of fuel they should use.'[44]

The government's policy was to bring coal supply and demand into balance. Alternative energy supplies also offered the long sought prize of 'avoidance of too great a dependence on deep-mined coal' and the NUM. Some 60 per cent of the NCB's costs were wages but national wage increases were invariably outpaced by local piece-rates so until the NCB could control its wage costs its position would worsen and more pits would close. Closures were inevitable but 'the high cost pits are situated chiefly in Scotland and Wales and special unemployment problems are created by the closure of pits owing to the fact that mining is carried on in close-knit communities.' Nevertheless, social dislocation did not justify preferential treatment for coal as 'I am sure we should avoid becoming dependent upon the last ton of coal from the deep mines with all that has meant in the past in rising costs and unbalanced labour/employer relations.'[45]

In December 1958 the NCB announced its first programme of pit closures. The closure of 36 short life pits would affect 13,000 mineworkers but as annual wastage was 60,000, redeployment was not likely to be a problem.[46] The NEC concluded 'little purpose would be served...simply opposing the Board's proposals' but noted the policy's result was 'whole villages becoming derelict in those outlying districts where...there was no alternative employment available.' A suggestion the NUM seek an immediate interview with the Cabinet 'to make it clear that the Union could not co-operate in the closure of pits which would result in unemployment' was rejected on the grounds that:

> any attempt to turn the problems of the mining industry into a political matter rather than an industrial one could be very serious for the Union...once Parliament was brought in or the Government asked to provide subsidies to the mining industry, this would be tantamount to giving them authority over the industry and over the jurisdiction of the NCB and the Union in the conduct of the affairs of the industry.

Consequently 'the *only* way out of the crisis was to give the maximum co-operation to the Coal Board with the object of ensuring that the steps which were ultimately taken would result in the least possible hardship to our members.'[47] '*The [NEC] had to face facts*' and utilise the experience gained in the earlier restructuring of the coke and by-product industry in which only 1,000 of the 12,000 effected had been made redundant thanks to close co-operation between the NUM and NCB over re-deployment.[48] The NUM expected nothing significant from Macmillan's government and its strategy depended on a Labour government being elected. In the meantime there was little the NUM could do stop closures because 'it is not the disposition and it is not the tradition of the British Trade Union and Labour Movement to fight political issues with the industrial weapon.'[49] The 1959 NUM Conference approved the NEC's policy but also heard Alex Moffat urge that Conference should 'indicate to the miners that they are to fight the policies that have been strangling the economy of this country'.[50] This was not to happen.

Whilst the NUM blamed a government induced recession for the industry's problems the NCB saw an irreversible shift in the pattern of fuel use taking place which required concentrating production on the most efficient pits and '*the elimination of that part of coal production which was highly uneconomic.*' This came at a time when NUM and NCB relations were at their worst since 1947 but the NUM never contemplated renegotiating the relationship as this 'was clearly a time for the Union and the Board to work together first to ensure the success of nationalisation and because the industry was in difficulties that affected both men and management.' A further reason for co-operation was the concern that the crisis might be used to justify claims that public ownership had failed and ought to be further reformed.[51] There were voices who opposed co-operation in the run-down. Jack Tighe (Notts) complained that 'We are sitting back at the present moment watching a Tory Government...gradually destroying that which has taken generations...to build up', while Jim Hammond described the 'closure of any pit for economic purposes or economic reasons [as] a complete departure in principle from the attitude this Union has held for many years'.[52] The NCB's analysis was economic, the NUM's political (the government's attack on full employment) and its solution was a planned economy, but this was hardly feasible under a Conservative government so to defend their members' interests the NEC agreed to co-operate with the NCB.[53] Not to do so would, it was argued, leave the NUM with no influence over the closures and meant abandoning their members. An industrial response raised the possibility of conflict with government so the NUM had no other option than accepting the NCB's approach. The April 1958 Special Conference did not question the fundamentals of the NCB's case and devoted most attention to ensuring that adequate consultation would take place before any closure and at the July Conference W.E. Jones, the NUM President, rejected unequivocally the proposition that 'industrial workers and the trade unions should use their unquestioned industrial and economic power in order to bring down the Government.' Instead trade unionists 'must place full reliance upon political democracy' and any failure to do so 'would reduce the possibility of an outstanding Labour Party success at the next and subsequent General Elections.'[54]

The NUM were determined to prevent any return to the 'harder work and less money' approach of the 1920s and 1930s. At CINC Horner had 'agreed fully that price cutting was suicidal. He preferred the alternative of a cut back in production and concentration on producing the coal required. *The Board should determine the amount of coal they could sell at economic prices*'. Any alternative to closures had been rejected by the Minister who had decreed 'that regard must be paid to consumer choice, and that if customers preferred oil to coal, this should be made available to them.'[55] The logical result of this was the NUM accepting closures on economic grounds whilst striving to ensure that their effects were minimised. Sam Watson expressed the NUM's dilemma succinctly: too much coal was being produced and 'no matter how much we shout and no matter how many speeches we make, you will not alter that fact.' The NUM must co-operate with the NCB to improve efficiency, recognise that coal did not enjoy a special dispensation from the laws of economics, and the NUM's first concern should be safeguarding those mineworkers who would remain in the industry.[56] The NUM had to avoid a repetition of the 1920s and 1930s: 'if you allow a recurrence of that depression in the coalfields – if it can be avoided and you fail to do it – then you are not worthy of the offices that you have been given by the men in this organisation.' Horner argued anything other than a joint approach to the industry's problems would destroy public ownership and 'so long as we have a Parliamentary system and the unions are subservient to Parliament, and unless we are prepared to challenge Parliament, then we have to try to change the policy of the country by changing the Government of the country.'[57] Alex Moffat doubted the joint approach would save any pits asking, 'is the policy of the NCB different from the old employers? Because after we have talked, is it not true that they close the uneconomic units just the same?'[58] Delegates agreed by 502 to 238 to co-operate with the NCB whilst seeking a commitment from the Labour Party for a coal-based fuel policy.

In May 1959 Percy Mills had reminded his colleagues that 'our present fuel policy is broadly based on freedom of choice for the consumer and equality of opportunity for the producer' whereas the NUM/Labour Party/TUC policy meant protecting the coal industry.[59] In 1957 the NCB had responded quickly to what it believed was a short-term crisis of overproduction but a piecemeal approach was no longer viable so 'we have to examine whether any degree of protection of the coal industry is justified, or whether we are facing an economic change which will have to be borne in the interests of efficiency and progress.' Mills believed the options were limited. He was concerned about any further increase in the heavy fuel-oil tax, open-cast mining was being run down despite its profitability, there was no point in stopping the building of oil fired power stations as they were practically finished and cutting back the nuclear power programme further would have little effect so 'If we are not prepared to implement, any further than we have done, the proposals put forward by the [NUM] and the Opposition, it seems inevitable that the coal industry must shrink in size.' The NCB was assuming a 200m ton production capacity by 1965 which implied the closure of 200 existing pits (100 by exhaustion) in five years. Mills argued it made sense for the government to wait until the Board had finished its appraisal but he wanted the Cabinet to reaffirm the policy of consumer choice. Cabinet discussed Mills' memorandum at length and agreed that 'If the policy of freedom of choice for the consumer were continued, the coal industry must contract.' This was tempered somewhat by Macmillan who, although in agreement with contraction, argued 'in the special circumstances of the coal industry, the government's fuel policy ought not to be based on too rigid an application of a policy of freedom of choice. It should be determined by practical considerations.'[60]

The NUM's reluctance to contemplate industrial action derived not just from the public ownership ethos, constitutional politics, or the legacy of 1926 but from a judgement the membership were unlikely to respond to a strike call. Ministers also knew this as it was manifest that the NUM would do nothing to destabilise the relationships built up since 1946-47. In 1957 the NUM Conference had approved a resolution calling on the NEC to examine the NUM's commitment to refer disputed pay claims to the NRT. Only in March 1959 did the Union's members on the JNNC meet to consider the resolution and they made it clear they supported no change in the conciliation procedures.[61] Faced by protests from Yorkshire and Scottish Areas at the NEC's inaction, Ernest Jones asked 'Do you think...this would be an opportune time to embark on something which even suggested strikes?' and, doubting that the men would obey a strike call, he concluded that 'We have had a lot of experience of strikes, and I can recall 1926, and by heavens we do not want to repeat that.'[62] As a result of the NRT's Thirty-Fifth Award in August 1960 and the resentment it caused some NEC members called for an overtime ban. This was not possible because of Clause 9 of the Conciliation agreement but all members of the NEC were unhappy with the Thirty-Fifth Award and it agreed that the union's JNNC members should review Clause 9 and report to the NEC. The NEC accepted a recommendation that the NUM withdraw from Clause 9 but warned the NCB that whether it did so or not depended on the Board's attitude to its 1960 wage claim. The NCB was not felt to be sufficiently accommodating and in July 1961 the automatic reference clause was amended on NUM insistence.[63] This change was symbolic as it led to no change in the NUM's behaviour.

Fundamental to the NEC's policy in the late 1950s was a determination to maintain the NUM's unity, a motivation emphasised by Paynter in his memoirs. The NEC believed that those most affected by closures would be most likely to strike and they would inevitably appeal for support from other coalfields, but those working in safe pits were thought much less likely to strike because they might place their own pits in jeopardy. Furthermore, if a strike was called and the NUM was defeated, pits would continue to shut, the Union's leaders would be blamed, again raising problems of internal unity. A lobbying and party-political campaign offered the best opportunity to maintain unity with the minimum of sacrifice and held the possibility of success.[64] The paradoxical effect of the NUM's political strategy was to depoliticise the mineworkers who 'were diverted from seeing their situation in a political context, with political causes and political solutions. They were encouraged to think there was no solution. The issue was converted into one of self-preservation, of looking after oneself and ones family.'[65]

The NUM's strategy depended on the election of a Labour government, a possibility eliminated by the Conservative victory in the October 1959 election. After the general election the Cabinet returned to the question of the coal industry. As we have seen, Richard Wood, Mills' successor, did not challenge the consumer choice policy but drew attention to the 'social problems' inherent in a large closure programme. Ministers agreed that 'It was right in the national interest, to refrain from protecting the coal industry from economic competition' whilst helping the industry to become more efficient through mechanisation. In the long-term 'coal would become an increasingly difficult and uneconomic source of power. Constant attention would have to be paid to the problems of redundancy in the areas most severely affected by the continuing contraction of the industry.'[66] The NUM recognised fuel policy would not change fundamentally and five more years of Conservative rule would lead to further closures. Paynter, who took over from Horner in the summer and who previously had called for a more 'active' policy, reiterated 'our general policy is directed towards securing political measures to secure [industrial] remedies.' Strikes would lose the NUM trade union allies, would jeopardise public support and bring the NUM into direct confrontation with government and Jones repeated the primacy of the party-political line as 'if we visualise any other sort of action, we are visualising bringing anarchy and chaos and hardship to the coalfield.'[67]

The appointment of Alf (later Lord) Robens as NCB chairman in early 1961 caused some trepidation in the NUM despite his service to the labour movement as union official, Labour MP and minister because he was known to favour de-centralisation.[68] An NEC sponsored Emergency Resolution to the 1960 Conference had restated the NUM's total opposition to any tinkering with the NCB's structure. Whilst decentralisation was not denationalisation Paynter argued it could have similar consequences because 'It can mean the revival of a price war between coal produced in the various coalfields...It could mean the development of disunity and disintegration within our own organisation because of the economic conflict that would arise between coalfield and coalfield.' Changes already underway at the NCB meant there would be greater political control and a proposed revised financial regime represented the transformation of the NCB into a state corporation which would provoke 'such bitter opposition and resistance as can only be paralleled with the struggle that took place in 1926.'[69] Moffat, seconding the resolution, argued the Conservatives were revealing once again their historic hatred for the miners and public ownership. He too warned of 'forceful resistance' from the mineworkers and that 'there is only one source left to the miners' Union and that is to use the industrial power and the strength of this organisation.'[70] The only significant dissent came from Sam Watson who urged the NUM have 'confidence in a man [Robens] who has spent his whole life in the working-class movement, has twice been a Cabinet minister and is also one of the ablest men we have'. He criticised those in the NUM who preferred 'a retired general and a retired admiral or a member of the Tory Party' as chairman.[71]

The NUM was deeply suspicious of Robens' appointment as rumours circulated in the industry that ministers were keen to extend the Fleck Report and promote competition between coalfields. Robens was known to favour a 'looser' NCB structure and his appointment was seen by some as an attempt to make these changes palatable to the NUM. In his memoirs Robens states categorically that ministers never urged him to decentralise the industry but he favoured the decentralisation of production decision making with wages negotiated nationally so as to keep control of costs.[72] Robens' appointment produced a noticeable hardening of the NCB's attitude. He told the JNNC in January 1962 that:

> The industry had to pull itself up by its own boot-straps – it was no good waiting and hoping for something to turn up. Whatever external forces may have been at work, the Industry had its own failings...Failure to achieve a great leap forward in productivity...would mean more price increases...The result of this would only be further contraction and more closures. The choice lay before the Industry.[73]

His robust approach to the industry's affairs worried Macmillan. Robens proposed to price Scottish coal out of the market to speed closures and Macmillan's dilemma was 'whether or not to agree with Lord Robens and the Coal Board, who want to raise the price of Scottish coal by 15s 0d

[75p] a ton. Scottish coal is run at a heavy loss, year after year. Lord Robens wants to *reduce* demand and so help in closing the worst pits.'[74] The NCB in general and Robens in particular, were reluctant to agree to a capital reconstruction which would reduce the Board's accumulated debt as they feared it would be a *de facto* declaration of bankruptcy and would reduce the pressure inside the industry to cut costs and uneconomic capacity.[75] Robens was successful in securing help for the industry from the Macmillan government and he praised Richard Wood for his compassion and understanding, and under him 'the coal industry got the most effective protection that ever came its way.'[76] Robens held informal talks with Selwyn Lloyd, the Chancellor of the Exchequer, before the 1961 budget which instituted a 2d per gallon tax on heavy fuel oil burnt in power stations. This was widely seen as protection for coal and it increased coal's competitiveness in electricity generation by £1 per ton.

In the discussions of the NCB document *Prospects for 1961*, the NUM again called for a joint NUM/NCB approach to government but the NCB refused absolutely to pursue a 'political' line. Bowman told Paynter that 'The Board would make their representations in their own way and the Union were, of course, free to take what action they pleased.'[77] The NEC's 'inner cabinet' recommended the NEC seek a meeting with the minister and then call a Special Conference to review the results and determine strategy.[78] The NUM met the minister for the second time in two years in November 1960. The NUM's position remained that ministers were biased against coal and that after 1947 the NCB had been forced by government to sell coal cheaply to subsidise industry. The government was responsible for the social consequences of this policy and was obliged to support financially the NCB, provide alternative employment in declining coalfields, reduce the CEGB's oil-burn, promote coal in gas manufacture, restrict industry's use of oil, and end opencast mining. However, 'the Minister's replies were all in the negative.' Wood's written response evoked the observation from the NEC's Economic Sub-Committee that 'the Minister's document adds nothing significant to his oral reply' and concluded there was no point in a further meeting.[79] At the instigation of the NEC Macmillan agreed to meet the TUC Production Committee to discuss the situation in the coal industry and the wider problems of fuel policy. This took place on 2 February 1961. Macmillan adhered firmly to established government policy but expressed sympathy with the NCB's difficulties.[80] A further meeting with the Minister in June 1961 was precipitated by the proposed import of US coking coal for the South Wales steel industry which the NUM contended broke the government's declared intention of addressing the balance of payments deficit and maximising indigenous fuel use. Wood responded that no government could block the importation of cheap fuel as this would boost the UK's competitiveness. The NUM reiterated its charge the government 'had no co-ordinated fuel policy [and] depended solely on market forces' to which Wood replied that government 'was somewhere in between the free-play of the market [and] protection...you could not force industry to use coal'.[81]

At this time ministers were preparing a White Paper *The Financial and Economic Obligations of the Nationalised Industries* (Cmnd.1337), which sought to bring them under closer ministerial supervision and make them more business like. The NUM warned that any moves towards either decentralisation or greater ministerial control 'would be resisted by the [NUM] with every legitimate means at its disposal.'[82] The NUM played no role in the negotiations between the Ministry and the NCB over the latter's financial obligations and the White Paper was first discussed at the NUM's request at the CINC in June 1961. The NUM's perception was that Cmnd.1337 sought to impose on public industry operating procedures comparable to private industry with a view to making a 'profit'. The NCB pooh-poohed this, arguing that the NCB's financial objective would be compatible with CINA and would take into account the industry's special circumstances. Paynter argued forcefully that Cmnd.1337 implied a tighter government–board relationship than had been accepted as normal since 1946, that the NUM's exclusion from the discussions reflected this new relationship, and lurking within Cmnd.1337 was the spectre of decentralisation. Paynter argued the industry's plight urgently required a financial reconstruction, which was strongly opposed by Robens, to ameliorate its accumulated debt and reduce interest payments to help it to compete in a rigged energy market. The Board disavowed any knowledge of future government plans and promised to inform CINC of further developments.[83] As 60 per cent of the industry's costs were wages the NUM feared too restrictive a financial objective would cost jobs and their fears increased when Robens revealed that discussions with ministers on a financial target had started but that little progress had been made. If agreement could not be reached Robens warned that the minister had the power to set the NCB's financial objective and there was no provision in the procedures for the NCB consulting the NUM via CINC so the best he could offer was that the NUM make its own representations to government.[84] The NUM regarded the restructuring of the NCB's finances as the key to coal's immediate problems. This, as we have seen. was resisted by Robens. The minister's intention had been to set the NCB's financial target by 1 January but had agreed to postpone this for twelve months as the Ministry of Fuel was engaged in a detailed review of the coal industry as part of an NEDC study into the preconditions for rapid economic growth. Robens stressed there was no suggestion of structural change as the industry would have a single financial obligation.[85]

The proposed financial objective was finally outlined to the NUM in October 1962. Its ultimate objective was to make the NCB self-financing but this would require substantial productivity increases and the speedy elimination of high-cost pits. The NCB had decided not to press for an early capital restructuring, a decision which bitterly disappointed the NUM but which had been accepted by the minister.[86] The Coal Industry Bill (1961) promised up to £50m by the end of 1962 to relieve the Board's accumulated deficit, a sum regarded by the NUM as totally inadequate, and the Conservative government (*not* the Board) was blamed by the NUM for the social consequences of inadequate support which would involve, according to the NUM, the virtual elimination of the Scottish coalfield. There was further confusion when the minister delivered a gloomy statement (23 October) about the industry's future, contradicting the NCB's optimism that a corner had been turned and that there was no need for further substantial closures. NUM suspicions were further roused by its exclusion from the discussions over the Board's financial objective. The NUM was in a difficult position as it had argued consistently that it had no responsibility for the NCB's balance sheet and its main financial concern was the accumulated deficit but this would change '[I]f government was to insist the industry made a profit this would require both huge price increases and pit closures both of which were not acceptable to the NUM'.[87]

The July 1961 Pay Pause and the subsequent 'guiding light' policy were a further factor in coal politics. Macmillan feared pay policy would provoke serious industrial and political problems and his government would have to face 'the danger of strikes and how to keep services going.'[88] Although the power workers broke through the policy in November one union Macmillan did not have to worry about was the NUM. The NUM did not defy the Pay Pause because the NUM did not want to damage any improvement in coal's prospects and it wished to avoid a clash with the Board and government while they were negotiating the Board's financial objective. Paynter argued that compared to other groups in the public and private sector the NUM had done relatively well, furthermore 'it is quite obvious that there is not a disposition inside the whole of the trade union movement' to challenge government policy.[89] Selwyn Lloyd, the

Chancellor of the Exchequer, believed that despite their public opposition the TUC would co-operate with the policy, but privately Macmillan was deeply concerned: 'We are now faced with the wage demands of the Coal Miners and the Railwaymen (as well as Gas, Bus and some other public employees).'[90] Un-minuted informal meetings between the NUM and NCB were held to resolve a range of unsettled disputes on pay and hours which had been complicated by the 'political side of the situation.'[91] The NUM's national officials met TUC representatives on 12 February to discuss the wider implications of the NUM's current wage claim which had been discussed previously by a special TUC Finance and General Purposes Committee. The TUC conceded government was indeed influencing public sector wage bargaining but the TUC had concluded that the diversity and complexity of public sector wage bargaining meant a co-ordinated response was not 'possible or desirable.' Whilst the NUM did not agree with the establishment of the NEDC it would, despite protests from South Wales co-operate with it so as to challenge government policy from the inside.[92]

Public sector pay remained a constant source of concern for the government. 'What authority', Macmillan mused, 'should the Government attempt to exercise over their chiefs whose legal position in relation to the central Government was delicate and obscure? Could a general directive be issued? Could a specific order be given? Or could we only appeal to a sense of patriotic duty?'[93] The problem was amplified by having to deal with such an independently-minded chairman as Robens. In late November 1961, for example, there was a row in the House of Commons when Wood implied that he had warned the NCB not to breach the pay policy but in February 1962 Robens offered the NUM a 4 per cent increase which was a breach of the 'guiding light' policy. In his memoirs Robens concedes he knew very well what government policy was but:

> I received no instructions from the Ministers or any other member of the Government. I regarded myself as being perfectly free and uncommitted in my approach to the current negotiations. I further declared that if I had been instructed as to my course of action, then I would have had to inform the [NUM] to transfer their negotiations to the Government.[94]

Robens, not the NUM, was the main source of the Macmillan government's anxiety about the coal industry.

In June 1962 the NUM met representatives of the Board of Trade and Ministry of Power to discuss the social consequences of pit closures. Ministers expressed regret over the social costs of closures but endorsed the policy concentrating on profitable pits and coalfields. Regional policy would be used to provide alternative employment but loss making pits would not be subsidised for social reasons and whilst the government did not wish to see unrestricted competition in the fuel economy, neither would it guarantee markets for coal.[95] The competitiveness and profitability of the industry had been reviewed by a group drawn from the Ministry of Power, Treasury, Board of Trade and NCB and concluded that current policy was broadly correct. This policy was that 'mines making no contribution to gross profits or depreciation' should be closed as quickly as possible but that the resulting social and political problems could not be solved solely by the NCB. The rundown in Scotland, South Wales and the North East was so severe and taking place so quickly Wood believed there was danger that this 'might be attributed to government policy.' Regional policy would offer some respite but it was unlikely to be fully effective because of the post-July 1961 recession which made businesses reluctant to move to development areas and there was a danger that the pit closures would result in a total collapse of business confidence in the affected areas. The Cabinet expressed concern that the NCB appeared more optimistic about the industry's future than did the departments concerned and noted that in 1961 162 pits had been defined as 'gross losers.' Concern was also expressed at the political impact of closures in Scotland but it was reported that many miners had speedily found work elsewhere so 'The general emphasis should be on dynamic action, and a more confident note should be struck'. The Cabinet felt that a capacity of 190m tons by 1965 was more realistic than the 200m ton target but concluded 'it would be inexpedient to appear to approve a specific target either for demand or output.' Finally, whilst the Cabinet agreed that the industry could not shoulder all the social costs of closures, it was unwilling to accept too great a responsibility as this might constitute a precedent.[96] In January 1963 Wood made a point of telling Robens about Macmillan's concern over high levels of regional unemployment, especially in the North East. Viscount Hailsham (Quintin Hogg) had been given special responsibility for the North East and Macmillan, who had sat for Stockton, a North East seat, between the wars, retained an emotional attachment to the area. Nevertheless, Robens was under considerable pressure to close pits at a faster rate, especially those which were thought not to have good medium-term prospects. Robens interpreted this as unwarranted ministerial interference which would provoke the NUM and disrupt the Board's plans and he refused. Fearful of the political consequences of Robens going public, Wood backed down.[97]

Faced by an invincible refusal to yield Paynter confessed that the NUM's efforts 'have not resulted in any change in policy or attitude by the present government, and I think we have to face the fact that as long as this Tory Government remains in power in Britain we are not likely to see the introduction of any effective policies that are directed to relieve us of our problems.'[98] There was, however, no need to despair. The TUC and the Labour Party were committed to a joint policy to preserve the coal industry, the Conservatives were increasingly unpopular and a revitalised Labour Party could well win the next general election.

1964: The Great Betrayal?

Between 1957 and 1969 the number of collieries fell from 822 to 299, a drop of 523 (63.6 per cent), manpower by 398,000 (56.6 per cent) and output by 67.6m tons (32.5 per cent). In the same period output per man year rose from 294.7 tons to 456.7 tons (an increase of 35.4 per cent), overall output per manshift from 24.9 cwts to 43.4 cwts (42.6 per cent) and the percentage of mechanised output increased from 23 per cent to 92.3 per cent. No other British industry underwent such a massive structural transformation and remarkably this loss of capacity and employment, and the attendant social dislocation, occurred without any resistance from the NUM. The NUM believed the incoming Labour government in 1964 was committed to avoiding this outcome. An important element in coal politics was the perception that the NUM had been 'betrayed' by the 1964-70 Labour government which reneged on a commitment made in opposition to guarantee the industry an output of 200 million tons a year by 1965. Betrayal was denied by ministers but contributed significantly to political change in the NUM at the end of the decade.

Harold Wilson enjoyed a close connection with the coal industry and the NUM. Wilson moved from an economics post at Oxford and become a wartime civil servant at the Mines Department and subsequently at the Ministry of Fuel and Power. Wilson's first major task was calculating coal output and in so doing he later claimed to have formulated for the first time the notion of productivity. Under Hugh Dalton and his deputy Hugh Gaitskell, Wilson's star rose as he moved from output statistics to miners' pay and he claimed a leading role in bringing the coal industry under state control in 1942. Wilson appreciated the value of a connection with the miners and he cemented this with his book *New Deal for Coal* (1945) which put the case for public ownership and it attracted the enthusiastic endorsement of Will Lawther. Wilson entered Parliament in 1945 and began a meteoric rise up the ministerial hierarchy; he was part of the team responsible for countering the 1947 fuel crisis. Wilson's reputation as a rising star was confirmed by his appointment as President of the Board of Trade. Wilson's resignation from the government in company with Aneurin Bevan over NHS charges did him little political harm. His supposed Bevanite links in the 1950s distanced him from the increasingly right-wing NUM leadership but his Northern origins worked in his favour and distinguished him from middle-class Labour intellectuals, he had access to the left in the NUM and his record as a minister in the Attlee government reassured suspicious minds. After 1957 until his election as Gaitskell's successor as party leader in 1963 Wilson was a key figure in opposition policy making on coal, chairing the Labour Party NEC's Fuel and Power Committee.[99]

The 1958 Labour Party Conference had approved unanimously an NUM resolution calling for a planned fuel policy with a guaranteed place for coal and in early 1959 the NEC's Home Policy Committee agreed on the need for a planned, comprehensive fuel policy.[100] However, formulating a policy was fraught with problems, one of the most serious was the suspicion of other fuel industry unions that their members' interests might be sacrificed to those of the NUM. This led to the formation of a Joint Committee on Fuel and Power Policy drawn from the party, the NUM and the TUC. The plight of the communities affected by closures was balanced by the difficulties of forecasting energy demand and a reluctance to encroach on consumer choice. Labour's *Challenge to Britain* (1953) had called for the recruitment of more miners and increased output; 'These seemed valid approaches to policy in 1953 – but in only six years the situation had changed radically.' So Labour was hesitant about being locked into a particular policy. A national fuel policy would be complex and require frequent revision so the committee recommended an interim statement to reassure the miners and a more detailed policy would then be prepared after consultation via the Joint Committee.[101]

The Joint Committee began by considering the NCB's *Revised Plan for Coal* which had envisaged a coal market in 1965 of 200-215 million tons and the closure of between 205 and 240 pits, 170 by exhaustion or merger. However, 'It cannot be emphasised too strongly that there are considerable hazards in forecasting the demand for coal' so:

> it would be best to assume that the NCB's 'range' of coal consumption is rather optimistic and it may be wiser to work on the assumption that inland coal

consumption will, at the most, be 190 million tons in 1965...this implies a total deepmined production of less than 190 million tons[...]

Assuming economic growth continued all accumulated coal stocks would disappear in the early 1960s and there would be a once and for all contraction in the number of pits with closures concentrated in the early 1960s.[102] This was broadly the outcome that the Macmillan government was aiming for. Having lost the 1959 election Labour hoped they would not have to deal with coal's decline and that the resulting opprobrium would fall on the Conservatives. The Joint Committee accepted with no discussion the inevitability of coal's decline and agreed what the industry needed was a 'cushion of time' to adapt to the new situation by, for example, ending open-cast mining and stopping the conversion of coal-fired power stations to oil burning. Ernest Jones, the NUM President conceded the irreversibility of the flood of cheap oil and made it clear the NUM was not arguing for the retention of all mines but for transitional support and a fuel policy to end uncertainty. The Joint Committee did discuss at great length the various energy projections for the mid-1960s and concluded there was no point in trying to second guess the NCB. However, the committee agreed that their object should be to secure a guaranteed tonnage for the NCB, a decision influenced by the views of E.F. Schumacher, the NCB's Chief Scientist. Schumacher argued against too rapid a rundown and supported a new policy to replace *Revised Plan For Coal* which would aim at a production of 200 million tons of coal by 1965. Uncertainty in energy prediction meant, Schumacher contended, that policy makers ought to err on the side of caution to ensure sufficient coal supplies.[103] From this point the Joint Committee was dominated by the size of coal's guaranteed tonnage.

The NUM disputed both government and NCB energy projections. Although it accepted a range of 180-190 million tons by 1965 as realistic, it was more concerned with securing a commitment that coal would remain the economy's basic fuel for the next 20-30 years in order to protect it from short-term fluctuations in the energy market. None of this aroused objections from the members of the Joint Committee which readily agreed coal would have priority in any fuel policy but this merely recognised reality. Even after the rundown coal would remain the country's largest single energy source and there was no competitor yet able to take over coal's base-load role in electricity generation. Furthermore, coal's primacy was not incompatible with a slimmer, more productive industry.[104]

Wilson's draft stated *inter alia* that the first priority of Labour's fuel policy was 'the use of indigenous coal' and '*the establishment of a guaranteed minimum for coal production, which the Government should make effective by all means within its power.*' This would provide a secure production floor and was based closely on a statement made by Robens, when still an MP in the House of Commons a year earlier.[105] This left the size of the production guarantee open and this omission was raised by a TUC anxious to reconcile all energy industry unions to the proposed policy. The TUC questioned the economic and political wisdom of a production floor for coal and as a result Wilson's paper (RD 63/June 1960) was modified. The Joint Committee agreed:

> the Government must immediately set a realistic minimum figure for coal production for some years ahead. This is the industry's major need. For, without a specific target, the [NCB] will not be able to plan sensibly its investment and production programme, nor will the crisis of confidence that those working in the industry are experiencing, be overcome.

This still left open the specific size of the production target and the report only stated that the Conservative government should set one. Doubts about the policy's wisdom were not confined to the TUC and were reflected in the statement that 'This target figure should be kept under constant review and adjusted according to changing trends in consumer preference, the growth of energy requirements, fuel and power technology and long-term changes in the balance of payments.'[106] These caveats were sufficient to neutralise the policy.

Neither of these paragraphs appeared in RD63. RD77 was passed to the Home Policy Sub-Committee which discussed it extensively and made further amendments. *Fuel and Power: An Immediate Policy* cited the Ridley Report to justify the need for long-term planning, rejected Conservatives' 'unregulated competition' and charged that the industry had been allowed to contract in an unplanned fashion producing a crisis of morale. A government endorsed minimum output target would provide stability and confidence but this would be reviewed periodically, oil usage and open cast mining would be curtailed, heavy fuel-oil taxed to discourage its use in power stations and coal stocks would be subsidised for social reasons. Except for the production target all of these measures were acceptable to the Macmillan government and the most sensitive part of RD77 was a future Labour government's policy:

> the Labour Party and the TUC do not propose that the present pattern of fuel consumption should be made permanent – still less that the Government should dictate to consumers the kind of fuel that they should use. But given the nature

of the fuel and power industries, it is imperative for the Government to ensure that adjustments are made gradually and that the long-term needs of the nation are not jeopardised.[107]

The report noted that the NUM had not asked for existing patterns of fuel use to be set in stone but only that the restructuring be phased and conducted sympathetically. In November the Home Policy Sub-Committee concluded the publication of *Fuel and Power: An Immediate Statement* meant 'the most urgent policy needs in this field have been met.'[108]

Fuel and Power: An Immediate Statement avoided specifying a target so where did the 200 million target come from? It flowed from two sources: first, the NUM. Here 200m tons took on enormous significance in the early-1960s as the pits closed, being wielded as a talisman against the Macmillan government. A second influence was the impossibility of accurately forecasting energy demand, especially as the economic slump after July 1961 was quickly replaced by the euphoria of the NEDC's 25 per cent growth over five years target and Labour's emerging National Plan both of which envisaged increased energy output. As part of economic policy making in Opposition the Labour Party NEC appointed an expert group to advise the both the Joint Committee and the Home Policy Sub-Committee on energy policy. Although *Fuel and Power: An Immediate Statement* remained the basis of Labour's energy policy the ideas of this group were to be of great political significance. The experts' view was that a 200m ton target was justified by economic growth, the state of the NCB's finances and the need to supply a range of markets. The investment of the 1950s was coming to fruition and the NCB were rapidly closing high cost pits, these developments would balance each other and should proceed in parallel. A faster rate of closures would create an imbalance, sterilise valuable coal reserves and risk non-cooperation from the mineworkers. Taking all these factors into account the group concluded that the break-even production point would be 180-190 million tons and economic and political prudence indicated a capacity of 200 million tons to be on the safe side in an uncertain climate.[109] The expert group was then called upon to produce a report on the scientific and economic aspects of fuel policy.

The Joint Committee revived its work on coal policy in July 1963, now chaired by the Deputy-Leader, George Brown. Richard Crossman, who was to bear the brunt of the 1967 pit closure crisis, and Fred Lee, Wilson's first Minister of Power and a vocal advocate of an accelerated pit closure programme, were also members. Building on Schumacher and Robie's paper the group recommended maintaining the coal industry at 200m tons at least in the short run:

Coal is an asset of a special kind, quite different from most other economic assets: it is non-renewable and inflexible. It cannot adjust itself quickly to changing market situations, and its current health and future availability depend on the cautious and deliberate pursuit of far-sighted policies of conservation. Its geological and sociological aspects cannot be preserved in any other way.[110]

This policy required a much higher level of ministerial intervention than that advocated by RD63 and the 1952 Ridley Report. Crossman welcomed RD480 (Revise) as 'politically, very acceptable' but expressed concern at the lack of firm evidence about future oil supplies (information controlled by the oil companies) and ominously Fred Lee 'felt that there was a case for a somewhat larger nuclear energy programme.'[111] RD480 effectively became Labour Party policy and the 200 million tons target was established when Brown wrote to the NUM's NEC detailing the content of the policy paper.[112]

Labour's 1964 manifesto *Let's Go with Labour!* promised 'We will have a co-ordinated policy for the major fuel industries' but said nothing specifically about coal. Coal's prospects to 1970 were thought reasonably good but in the summer before the 1964 election the Joint Committee was warned 'that the discovery of very large quantities of low cost gas [in the North Sea] would inevitably have major repercussions on the market for other fuels, particularly coal, and would cause major disruption unless its coming into use was planned within the framework of a national fuel policy.' Wilson had noted in June 1960 that if its introduction was not planned 'we may see coal – as in cotton but on an immeasurably vaster scale and to an extent which imperils our industrial future – run-down to such a point of the industry that shortages once again develop, and the industry is unable to grasp the opportunities which are awaiting it.'[113]

Labour's Accelerated Closure Programme

The NUM expected that a Labour government would revive the fortunes of the coal industry by implementing what the union regarded as a commitment to a 200 million ton coal industry. They were to be sadly disappointed. On 12 November 1964 the NUM's three National Officials met the Minister of Power, Fred Lee, at his request. Lee told them the ministry had not yet had time to prepare the energy-use projections on which Labour's fuel policy would be based but 'there was no question that the Labour Government would stand by the assurances which had previously been given on behalf of the Labour Party.' The National

Officials were told that fuel policy would be based on planning and co-operation, not competition between the fuel industries, violent fluctuations in demand for coal would be smoothed by public sector consumption and stockpiling. Labour's policy of planned economic growth would sustain energy demand.[114]

This aroused NEC suspicions and the Economic Sub-Committee sought a second meeting with Lee where they concentrated on securing a public commitment confirming the 200 million ton target. Lee's response was that 'his outlook was "pro-coal"' but that 'it was no good looking at a Fuel Policy against the demands in 1965' and while the NUM's views would be taken into account Lee 'could not, however, pre-judge these decisions or say when precisely they would be reached but he and his colleagues were determined to do so as soon as practicable.' Rumours were rife of a major expansion of oil refining capacity and that in response the coal industry would be cut to 160-180 million tons. Morale in the pits was low and skilled manpower was leaving the industry to such an extent that production might soon be jeopardised and the NEC suggested the government halt pit closures (except of grounds of exhaustion) until its fuel policy was finalised. The Miners' Parliamentary Group was requested to use its influence and a second meeting was sought with the Minister of Power and Minister of Economic Affairs, George Brown, who was responsible for economic planning.[115] In response to a parliamentary question Lee insisted 'There is no question of my requiring an arbitrary reduction in coal output' but this implied a planned reduction. He continued, 'The objective of a coal market of about 200m tons will be sympathetically considered in the formulation of our long-term national fuel policy which will have regard to all relevant considerations.'[116] One week later in response to a request that closures be suspended Lee replied that 'the constitutional position was that the Board was generally responsible for the management of the coal industry by virtue of the nationalisation statute and that he, as Minister, had no power to intervene in decisions affecting the closure of individual collieries.'[117] On 9 February Lee repeated that 'The government accept for the present the case for trying to maintain the position of coal...at around its recent level of 190-200 million tons...I hope this statement of the government's intentions towards the coal industry will dispose of recent alarmist reports in the Press about the prospects of employment in the industry.'[118] NUM suspicions were confirmed by Lee's statement that 'About half the Board's output comes from profitable pits and half from pits that are making losses.'[119]

Oil was cheaper than coal and Ministers could not intervene with consumer choice. The NEC had already concluded government policy was 'not in line with the general understanding it had with the Labour movement. Deep concern was expressed at the continued reluctance of the Minister to commit himself to a definite output target for the industry' and with NCB support, Lee argued vociferously against a subsidy.[120] The Cabinet agreed there would no subsidy as this would be used to sustain uneconomic pits and the financial package was 'to provide special funds to speed the disappearance of uneconomic collieries.' Once the restructuring had taken place the result would be 'a healthy and viable coal industry capable of providing funds for further colliery investment from its own earnings.' When questioned on the future size of the industry Lee stonewalled making the familiar criticisms of the previous Conservative government's policy and arguing that Labour's policy was based on a consensus with the NCB 'and we believe that this provision will enable it [the NCB] to make the necessary closures in certain areas.'[121]

Sidney Ford, the NUM President, was a totally committed Labour supporter, but acknowledged 'we are faced with the prospect of a continuing indiscriminate contraction' of the industry despite undertakings given to the NUM. How should the NUM react? Ford urged delegates to remember that:

> For 13 years...we had a Tory Government. We must be constantly asking ourselves the question – do we want this Government to survive and succeed or are we prepared to seriously risk another 13 years of Tory Government? There can be no rocking the boat. This is as true of individual trade unions as of individual Labour MPs. There will be no forgiveness for anyone or any group who willingly helps to bring the Government down. We shall sometimes have to subordinate our own interests to those of the great majority.[122]

Even allowing for the enthusiasm aroused by the election of a Labour government with a narrow majority this was a remarkable statement. It was delivered only four days after Lee's statement in the Commons and as Lee was to speak at the NUM Conference, Ford's statement was interpreted as a pre-emptive strike against any opposition to government policy. Fred Lee's speech to the NUM Conference reiterated his Commons statement and coupled lavish praise for the miners with a warning that coal's future depended on removing 'the dead weight of many collieries which do not cover even their running expenses.'[123] Conference adjourned immediately to permit delegates to absorb the implications of Lee's statement which was more or less repeated by Robens when he addressed the conference as NCB chairman:

> The world does not feel that it owes us a living and consumers cannot be forced to buy a fuel they do not want. The market is a harsh employer and will dismiss even the most faithful servants without a moment's regret or a penny of severance pay. All the buyer is concerned with, is whether there is an alternative fuel at a lower price.[124]

In a speech which many think was one his best, Paynter reminded delegates that the industry's problems 'are not a consequences of actions that have been taken by a Labour Government. They are the consequence of accumulated actions and neglect for which the Tory government are responsible'. Coal would not get a better deal from a Conservative government and although Paynter agreed Labour had shifted ground after 1964 he believed the situation was not yet lost irretrievably and 'we have to remind the government that there is some political obligation on their part to us.' Equally, there were limits to the NUM's loyalty.[125]

On the afternoon of 8 July the NEC tabled an Emergency Resolution which questioned the wisdom of the accelerated closure programme. Moving the resolution on behalf of the NEC Abe Moffat hinted at the impatience of some delegates at Labour's refusal to implement the 200 million ton policy and Peter Heathfield, the future General Secretary at his first Annual Conference, cut right to the substance of the emerging controversy; 'I am very impressed by some of the very fine speeches and statements that have been made at previous Conferences but if one looks at events of the past few years, it seems that those high-sounding speeches and fine words were meaningless. They have closed pits.'[126] The Emergency Resolution was approved unanimously.

Manpower was flowing from the industry and the NCB's Marketing Department reported an accelerating loss of markets. At CINC Schumacher explored at great length the disparity between the Board and government's assessment of the industry and concluded that the Ministry's emphasis on one measure, production costs, would generate a spiral of decline. Jim Bullock of the British Association of Colliery Managers raised the question of the social consequences of the closure policy and Paynter disputed Schumacher's assumption that no further help would come from the government. Schumacher's response was that 'gross losers' would have to be closed if the industry was to have a chance and reiterated Robens' insistence 'no Government could issue a "blank cheque" and guarantee the sale of any given quality of coal. A national fuel policy would only amount to support which, with the proper effort, should enable the industry's aims to be achieved'. At the end of the meeting this exchange took place:

> Mr Bullock asked if the Board had acquiesced to the Government plan to cut down the size of the industry.
>
> Sir Humphrey Browne stressed that Lord Robens had tried as vigorously as possible to get the industry's target at 200 million tons a year but it was becoming clear that the Government would not agree to do so.[127]

On the following day the NEC discussed press speculation of a further rapid contraction of the coal industry as a result of the fuel policy White Paper which was under preparation. Lee denied this but the NEC interpreted the speculation as the result of government inspired leaks designed to 'soften-up' the NUM. Robens had already agreed to the acceleration of closures which would be concentrated in Fife, South Northumberland, West Durham, Cumberland and South Wales and as a result 'employment in coalmining would virtually come to an end in many of these districts.' Robens insisted that ministers had to shoulder their responsibility for the social costs and he pointedly asked for the government's full support in carrying out the closures.[128]

Labour's flagship policy, *The National Plan*, declared previous estimates of a coal consumption of 190-200m tons were unrealistic and estimated demand in 1970 at 170-180m tons despite a projected increase in energy consumption. Some NEC members concluded optimistically this was a projection not a target and coal could win a larger share but it was emphasised that fuel policy could not depart from *The National Plan*. The special measures to help coal adjust would continue.[129] By 1970 the industry would shrink to 170 million tons and 235,000 men with output concentrated on the Central Coalfields and that 'the limit of Government assistance had been reached, at least for the next five years.' The NEC's Economic Sub-Committee accepted that 'The ultimate size of the industry would depend on its ability to sell in direct competition with other fuels...the price of coal was of prime importance. The industry must get its costs down, a responsibility of both management and the Union.'[130] Robens announced that the industry would now plan on an output of 170-180 million tons by 1970, the industry could possibly achieve the 200 million tons even more rapid closures were needed to bring average costs down. Ford described *The National Plan* as 'disappointing', regretting the government's conclusion 'that it had met its obligations and that the industry would now stand or fall by its own efforts.' He warned critics of the policy that the NUM had no trade union allies and 'Neither the public, the TUC nor Labour Party was interested'. Paynter was more vehement arguing for 'continued pressure' on the Government and the suspension of all closures. He was supported by other NUM members on CINC who condemned the fact that 'Whole areas were being made derelict through closures. Durham had the best record in the industry for disputes and absenteeism but if closures continued this record of co-operation was in danger.'[131] It remained to be seen how mineworkers would react to this accelerated pit closure programme.

Notes

[1] *NUM (ACR) 1956*, pp. 87-8.
[2] *NUM (ACR) 1956*, p. 105.
[3] Aubrey Jones to Eden, 25 July 1956. *PREM 11/1230*.
[4] H. Macmillan, *Riding the Storm, 1956-1959*, Macmillan 1971, p. 30.
[5] *NUM (EC)*, 24 October 1956 (afternoon meeting).
[6] *JNNC (Union's Side)*, 20 December 1956, and *NUM (EC)*, 10 January 1957.
[7] *The Times*, 1 March 1957. The NEC reiterated its opposition to this appointment but decided nothing would be gained by continued protest.
[8] NUM, *Report of a Special Conference*, 25 January 1957, p. 20.
[9] *Report of a Special Conference*, 25 January 1957, p. 25.
[10] *Report of a Special Conference*, 25 January 1957, p. 28.
[11] *Report of a Special Conference*, 25 January 1957, p. 31.
[12] NUM, *Report of a Special Conference*, 1 March 1957, p. 9 and p. 11. Emphasis added.
[13] The Bonus Disqualification was an agreement whereby if a miner worked five full shifts he was paid for six, if one of the five shifts was missed for any reason the bonus was not paid.
[14] Meeting with the NCB on Wages, *NUM (EC)*, 11 April and 18 April 1957. Emphasis added.
[15] NUM, *Report of a Special Conference*, 10 May 1957, p.5 (Horner), pp.18-19 (Kane), and p. 20 (Main).
[16] *NUM (ACR) 1957*, p. 85 and p. 89.
[17] *NUM (ACR) 1957*, pp. 129-30.
[18] Joint Meeting with the NCB, *NUM (EC)*, 25 March 1958.
[19] Situation in the Industry, *NUM (EC)*, 4 July 1958.
[20] *NUM (ACR) 1958*, p. 27.
[21] *NUM (ACR) 1958*, p. 94.
[22] *NUM (ACR) 1958*, p. 96.
[23] *NUM (ACR) 1958*, p. 133.
[24] *NUM (ACR) 1958*, p. 136.
[25] *NUM (ACR) 1958*, p. 140.
[26] *NUM (ACR) 1958*, p. 142.
[27] *NUM (ACR) 1958*, p. 143.

28 *NUM (ACR) 1958*, p. 145.
29 NUM, *Special Conference Report*, 10 October 1958, p. 12.
30 *NUM (EC)*, 6 November 1958. Emphasis added. A compromise favourable to the NCB was reached as the focus moved away from pay to pit closures.
31 Macmillan, *Riding the Storm*, pp. 361-62.
32 *NUM (ACR) 1958*, p. 96.
33 *NUM (ACR) 1958*, pp. 135-36.
34 Situation in the Industry, *NUM (EC)*, 3 December 1958.
35 *NUM (ACR) 1959*, pp. 327-28.
36 *HC.187-I*, Report of the Select Committee on Nationalised Industries (Reports and Accounts), para. 118 (p. xxii).
37 *JNNC (122)*, and *NUM (EC)*, 19 February 1958.
38 C (58) 203, 30 September 1958. Fuel and Steel. The Effects of the Recession. Memorandum by the Minister of Power. *CAB129/95*. Lord (Percy) Mills replaced Aubrey Jones in January 1957. He was a close friend of Macmillan but was not in the Cabinet, a demotion which reflects coal's declining political importance.
39 *Report of the Committee on National Policy for the Use of Fuel and Power Resources.* Cmnd.8647 (September 1952), para 1 and para 4. Hereafter, *Ridley Report*.
40 *Ridley Report*, para 222(d) and para 223.
41 *5s H.C. Debs 614*, 23 November 1959, cols. 46-47. Richard Wood was appointed minister and to the Cabinet on 14 October 1959 and retained this post until 20 October 1963 when he was replaced by Fred Errol.
42 *5s HC Debs 612*, 3 November 1959, col. 977. Ending opencast mining was opposed by the NCB as these operations highly profitable. The electricity industry was keen to make more use of oil but as it was the NCB's biggest market to do so would worsen the crisis in the coal industry. The result was the compromise policy of dual-fired power stations which could burn either coal or oil which emerged in 1959 from a Ministry working group. See *POWE 37/555* for detail.
43 Situation in the Industry, *NUM (EC)*, 5 November 1959.
44 Appendix 1. Report of a Deputation to the Minister of Power, 18 November 1959. *NUM (EC)*, 19 November 1959.
45 C (58) 258, 19 December 1958. Fuel Policy and Problems. Memorandum by the Minister of Power. *CAB129/95*.
46 Ashworth, *The History of the British Coal Industry*, pp. 258-59.
47 Situation in the Industry, *NUM (EC)*, 3 December 1958. Emphasis added.
48 Situation in the Industry, *NUM (EC)*, 12 December 1958 (Meeting with the NCB). Emphasis added. Closures were implemented in Durham with few problems but in Scotland and South Wales there was unrest and in January a delegation of 200 South Wales miners and in February a Scottish delegation lobbied the NEC.
49 Situation in the Industry and Proposals for a National Fuel Policy, NUM, *Report of a Special Conference*, 27 February 1959, pp. 18-19. Jones attacked opponents of this policy as politically motivated (Communist). Responding to a question about what the NUM would do if the Conservatives were re-elected Jones stated the NUM would stand 'by the power of the ballot boxes' (p. 22).
50 *NUM (ACR) 1958*, p. 328.
51 Situation in the Industry, *NUM (EC)*, 18-19 March 1958 (my emphasis) and NUM, *Report of a Special Conference*, 11 April 1958, p. 14. The NCB identified 167 pits which were making a loss of £1 per ton.
52 *Report of a Special Conference*, p. 15 and p. 20.

53 Joint Meeting with the NCB, *NUM (EC)*, 25 March 1958.
54 *NUM (ACR) 1958*, p. 27.
55 *CINC (75)*, 13 May 1958, and Situation in the Industry, *NUM (EC)*, 4 July 1958. Emphasis added.
56 *NUM (ACR) 1958*, pp. 170-71.
57 NUM, *Report of a Special Conference*, 10 October 1958, p. 10 and p. 13. In November 1958 the NUM donated £35,000 to the Labour Party's General Election Fighting Fund.
58 *Report of a Special Conference*, p. 20.
59 C (59) 90, 26 May 1959. Fuel Policy. The Coal Problem. Memorandum by the Minister of Power. *CAB129/97*.
60 CC (59) 32nd Conclusions, 28 May 1959. *CAB128/33*.
61 *JNNC (Unions' Representatives)*, 23 March 1958.
62 *NUM (ACR) 1958*, p. 255 and p. 256. Yorkshire's reference back of the NEC's policy on the NRT was defeated by the relatively narrow margin of 408 to 307.
63 Wages – 35th Award of the NRT, *NUM (EC)*, 18 August 1960, *JNNC (137)*, 30 November 1960, and NCB, *Annual Reports & Accounts 1961*, p. 29.
64 W. Paynter, *My Generation*, G. Allen & Unwin 1972, p. 142.
65 V. Allen, *The Militancy of British Miners*, The Moor Press 1981, p. 65.
66 C (59) 180, 7 December 1959. Policy for Fuel. Memorandum by the Minister of Fuel and Power. *CAB129/99*, and CC (59) 62nd Conclusions, 10 December 1959. *CAB128/33*.
67 NUM, *Report of a Special Conference*, 20 November 1959, p. 15.
68 After a career as a union official in the Shopworker's Union, Robens was MP for Wansbeck (1945-50) and Blyth (1950-60). In 1946 he was appointed Parliamentary Private Secretary to the Minister of Transport, Parliamentary Secretary at the Ministry of Fuel and Power (October 1947-April 1951), and Minister of Labour and National Service (April-October 1951). In Opposition he was variously spokesman on Foreign Affairs, Fuel and Power, and Labour.
69 Emergency Resolution, *NUM(ACR) 1960*, p. 167 and p. 169.
70 *NUM (ACR) 1960*, p.171.
71 *NUM (ACR) 1960*, p. 171.
72 Lord Robens, *Ten Year Stint*, Cassell 1971, pp. 5-6. Robens made this position clear at a meeting with Butler in May 1960 where he offered the NCB chairmanship (pp. 7-8). Decline meant there would eventually have to be a major reorganisation of the NCB's regional structure.
73 *JNNC (141)*, 17 January 1962.
74 H. Macmillan, *At the End of the Day 1961-1963*, Macmillan 1973, p. 43. Diary for 26 October 1961.
75 Ashworth, *The History of the British Coal Industry*, p. 277
76 Robens, *Ten Year Stint*, p. 151.
77 *CINC (85)*, 20 September 1960.
78 Situation in the Industry, *NUM (EC)*, 20 October 1960.
79 Appendix I. Report of a Deputation to the Minister of Power, 2 November, *NUM (EC)*, 9 November, and Appendix II. Economic Sub-Committee, 24 November, *NUM (EC)*, 8 December 1960.
80 Situation in the Industry, *NUM (EC)*, 2 February 1961. See 5s *H.C. Debs 635*, 20 February 1961, cols. 51-53 for Wood's presentation of the principles underlying fuel policy.

81 Appendix I. Report of a Deputation to the Minister of Power, 13 June 1961. *NUM (EC)*, 30 June 1961.
82 Appointment of Chairman of NCB, *NUM (EC)*, 16 June 1960.
83 *CINC (88)*, 20 June 1961.
84 *CINC (90)*, 13 December 1961.
85 *CINC (91)*, 3 May 1962.
86 *CINC (94)*, 9 October 1962.
87 Situation in the Industry, *NUM (EC)*, 14 December 1961.
88 Macmillan, *At the End of the Day*, p. 37. Diary entry for 21 September 1961.
89 NUM, *Report of a Special Conference*, 23 February 1962, p. 17.
90 Macmillan, *At the End of the Day*, p. 62. Diary entry 15 January 1962. Wage pressure and a growing sense of relative economic decline led the Macmillan government to establish the National Economic Development Council (NEDC) in 1962 in an effort to forge a low-inflation, high growth consensus embracing government, employers and union.
91 Wages and Hours of Work, *NUM (EC)*, 1 February 1962.
92 Wages, *NUM (EC)*, 22 February 1962.
93 Macmillan, *At the End of the Day*, pp. 44-45.
94 Robens, *Ten Year Stint*, p. 46.
95 Appendix I. Report of a Deputation to the Minister of Power, *NUM (EC)*, 21 June 1962.
96 C (62) 104, 3 July 1962. Reappraisal of the Coal Industry. Memorandum by the Chancellor of the Exchequer. *CAB129/110*. CC (62) 44th Conclusions, 5 July 1962. *CAB 128/36*.
97 Notes of a Meeting between the Minister of Power and Lord Robens, 21 January and 30 September 1963. *POWE 52/4*.
98 *NUM (ACR) 1963*, p. 245.
99 B. Pimlott, *Harold Wilson*, Harper Collins 1992, pp. 77-91.
100 The Situation in the Industry and Proposals for a National Fuel Policy, *NUM Report of a Special Conference*, 27 February and *Proposals for a National Fuel Policy* (March 1959) formed the basis of NUM policy. See also Labour Party National Executive Committee. *Home Policy Committee Minutes*, 9 February and 9 March 1959. National Museum of Labour History, Manchester. Hereafter, HPC.
101 *RD 570/June 1959*. National Fuel Policy, and *HPC*, 29 June 1959. The interim statement was issued on 6 July.
102 *RD 10/December 1959*. National Fuel Policy, pp. 6-7. Joint Committee on Fuel and Power Policy Minutes and Papers, 13 December 1959-24 July 1963. Hereafter, *JCFPP*.
103 *JCFPP*, 15 December 1959, and Report of an Interview with Mr E.F. Schumacher, 12 December 1959.
104 *RD 22/February 1960*. Problems Arising from the Reduction in Coal Mining Manpower, *RD 26/February 1960*. Memorandum Submitted by the National Union of Mineworkers, and *JCFPP*, 11 March 1960.
105 *RD 63/June 1960*. Fuel Policy. Note by the Chairman (original emphasis), and *5s H.C. Debs 605*, 4 May 1959, col. 40.
106 *RD 77/July 1960*. Fuel and Power. An Immediate Policy, para 12-13. The TUC's reservations can be found in *RD 64/June 1960*. Suggestions on Content of a Longer Period Policy Statement.
107 *RD 77/July 1960*. Fuel and Power: An Immediate Policy, and *HPSC*, 20 June and 11 July 1960 which approved the amended document.

[108] *RD 90/November 1960.* Programme of Work. The Joint Committee was revived in 1962 to work on the impact of EEC membership on the coal industry.
[109] E.F. Schumacher and G.M. Robie, *The Size of the Coal Industry*, 12 September 1962 was prepared for the JCFPP and HPC.
[110] *RD 480/June 1963 (Revised)*. Fuel and Power. Report of Study Group under the chairmanship of Professor Newitt on the Scientific and Economic Aspects of a Long-Term Fuel Policy.
[111] *JCFPP*, 24 July 1963.
[112] G. Brown to H. Wilson, 5 February 1965. *PREM 13/328*. There are several scattered references to this letter in government and NUM sources but I have been unable to locate a copy to verify its contents. Politically, however, its contents are less important than its acceptance by the NUM as a commitment to a production floor of 200 million tons.
[113] *RD 808/August 1964.* North Sea Oil and Gas, and *RD 63/June 1960 (Revised)*. Fuel Policy. Note by the Chairman, para 27, p. 26.
[114] F. Lee, Coal Policy, 22 June 1965. *PREM 13/328*, and Situation in the Industry, *NUM (EC)*, 12 November 1964.
[115] Appendix II. Report of a Meeting of the Economic Sub-Committee with the Minister of Power, 7 January, and Situation in the Industry, *NUM (EC)*, 14 January 1965.
[116] *5s H.C. Debs 705*, 21 January 1965, col. 125. Lee repeated this on 1 February (col. 236).
[117] *NUM (EC) Special Meeting*, 4 February 1965. There was also a long discussion between the NEC and the NCB at this meeting on the industry's future which revealed considerable common ground.
[118] *5s H.C. Debs 706*, 9 February 1965, col. 52.
[119] *5s H.C. Debs 713*, 25 May 1965, col.221. Lee denied this implied half the pits should close.
[120] Situation in the Industry, *NUM (EC)*, 8 April 1965, NUM, *Report of the National Executive Committee*, May 1965, pp. 6-7, and R.H.S. Crossman, *The Diaries of a Cabinet Minister. Volume 1. Minister of Housing 1964-66*, Hamish Hamilton/Jonathan Cape 1975, p. 258. Entry for 28 June 1965.
[121] *5s H.C. Debs 715*, 1 July 1965, cols. 838-40, Crossman, *Diaries Volume 1*, p. 263 1 July 1965, and CC(65) 35th Conclusions, 1 July 1965. *CAB 128/39*. Mining MPs welcomed the package and one, Ness Edwards, described it as the best news for years.
[122] *NUM (ACR) 1965*, p. 84.
[123] *NUM (ACR) 1965*, p. 186. Will Whitehead (South Wales/NEC) insisted all delegates be provided with a copy of Lee's speech before the subsequent debate on the Situation in the Industry.
[124] *NUM (ACR) 1965*, p. 202.
[125] *NUM (ACR) 1965*, pp. 212-19.
[126] *NUM (ACR) 1965*, pp. 224-47 and p. 231.
[127] *CINC (109)*, 27 July 1965, and Robens to Lee, 27 August 1965. *PREM 13/1610*.
[128] Situation in the Industry, *NUM (EC)*, 28 July 1965. Robens feared that ministers anxious to avoid the blame for the closures would try and shift all the responsibility onto the NCB. Robens wanted ministerial hands to be dipped in the industry's blood.
[129] Economic, *NUM (EC)*, 16 September 1965. These special measures were: a continued ban on coal imports, a write-down of the industry's capital liabilities, continuation of the tax on heavy fuel oil, preference for coal in heating public buildings, and increased government support for the social consequences of pit closures.

[130] Appendix III. Economic Sub-Committee, 15 September, *NUM (EC)*, 16 September 1965, and F. Lee to D. Jay, 15 July 1965. *PREM 13/923*.

[131] The National Plan and the Coal Industry, *CINC (111)*, 18 October 1965, and G. Brown to H. Wilson, 18 October 1965. *PREM 13/923*. This correspondence stresses the importance of the correct presentation of the closures to minimise opposition in the Labour Party and amongst the unions.

Chapter 6

The Miners and Mr Wilson's New Britain

Introduction

The NUM's relationship with Harold Wilson's New Labour demonstrates the tensions and contradictions at the heart of British social democracy. The pit closure programme demonstrated the relative powerlessness of the NUM in Labour's internal politics because of history, loyalty, and a judgement that any Labour government was better than a Conservative government. Even if the NUM had been more assertive there is no evidence that the Labour government would have responded. The closures illustrated the tension between Labour's modernisation ethos and the problems modernisation created for the coal-mining communities. The NUM was reticent and defensive, afflicted by a deep inhibition against challenging an elected government which, in turn, reinforced the NUM's commitment to parliamentary politics and lobbying. The NUM had insufficient political resources to challenge the dominant civil service view of the coal industry. The NUM was actively excluded from core decision making on energy policy by a policy network which was pro-nuclear despite coal's centrality to electricity generation, and passively excluded from core decision making in the NCB by the industry's consultation and conciliation structures.

The 1965 Pit Closure Crisis

'This whole business of pit closures, far more drastic than those proposed by the Tories,' Crossman noted in his diary, 'is one of the most embarrassing duties our Government has assumed. And with some interest I watch Fred Lee trying to grapple with it. Certainly he is handling the miners far more savagely than I would ever dare to do.'[1] During the preparation of the *National Plan* George Brown had been convinced the economy would need more energy to achieve its growth targets but that these inputs should not necessarily come from coal, so the 200m ton industry was neither desirable nor sustainable.[2] The crisis broke after the publication of the White Paper, *Fuel Policy* (Cmnd.2798) in October 1965.

The NEC postponed its scheduled meeting of 18 November because Divisional Coal Board meetings on that day were to announce the scale of the accelerated pit closure programme. The NCB projected the closure or merger of 150 Category C (short-life) pits, 95 due to exhaustion and 55 for economic reasons, 81 were Category B (with a doubtful future) and 281 were Category A and would continue in production. The closures would affect 120,000 NCB employees of which 60,000 could be absorbed in pits within travelling distance, 25,000 could be redeployed to coalfields with a labour shortage, and natural wastage would account for many of the rest so, relatively few mineworkers would be unemployed. Ministers had agreed to pay 50 per cent of the social costs of the programme to a maximum of £30 million by 1970-1971. A minority on the NEC openly described the programme as a betrayal which should be resisted by all available means but the majority argued the NUM's duty to its members lay in ameliorating the effects of the closures and continuing to support the modernisation of the industry.[3]

On 25 November 1965 the Second Reading of Coal Industry Bill took place and Lee announced the acceleration of closures. He forecast 200 pit closures over the next five years, with half in the next two years, and 'the Coal Board aims to close all those collieries which fail to cover their running costs and are unlikely to do so in a foreseeable period of time.'[4] Tom Swain (NUM sponsored MP for Derbyshire North East) was bitterly critical of Lee and the government:

> I did not think that I should be speaking in a debate in 1965 on the introduction of a Bill which is the requiem mass on the mining industry. When I pulled the string which raised the [NCB's] flag at my pit head on 1st January 1947, I was the proudest man alive, not only because we had achieved nationalisation, but because Socialism had achieved nationalisation. I never thought that I would find a Minister of Power on the Socialist benches introducing a Bill of this character. He has completed the work which the saboteurs opposite have been trying to carry on for 13 years.[5]

Lee promised the industry aid 'on a truly massive scale' but the 200 million ton target was dismissed as not feasible 'short of totalitarian measures' which would distort the whole economy and encourage similar demands from other declining industries.[6] Robens concluded that 'It was evident that the political boundaries within which the industry must work had been set at least until 1970. He doubted that anything more than had already been promised could be obtained from the Government, it was up to the industry, in that context, to do all it could to help itself.'[7] Robens was angered by the accelerated closure programme which he believed jeopardised the creation of the modern, productive industry sought by the NCB, the NUM and government. Ministers were puzzled by Robens' response as they believed he would welcome the accelerated programme because it would achieve their common goal in a shorter time.[8]

Discussions with ministers continued into the New Year. A meeting with Lee and Douglas Jay and Lord Lindgren (both of the Board of Trade) began with Lee insisting the government had not 'betrayed the miners, [he] strongly refuted the charge that the government had not fulfilled the promises and pledges which had been made before the General Election [and] no one yet had clearly set out what was in their minds when they referred to a National Fuel Policy'. The NEC insisted that:

> the NCB was influenced by Cabinet decisions, and therefore the Board was in fact operating under the direction of the Government. The policy of the present Government was based on the premise that the future of the industry must be determined by freedom of consumer choice and free-for-all competition. It was suggested that the government had no National Fuel Plan, but in the main was simply projecting the policies which had been in existence under the previous Government.

The dryness of the formal minutes cannot disguise the NUM's frustration:

> There was no point in arguing with the Ministers as to whether or not the present Government had broken faith with the miners, but it was the common belief that the views of certain officials of the Ministry of Power under the Tory Government were in fact now present Government policy. The Government was therefore either guilty of political chicanery or had been hopelessly incompetent in the assumptions they had made before the election. Government policy accepted free-for-all competition, and in free-for-all competition price was of paramount importance. The ministers agreed 'political decisions' had been taken which directly affected coal but pit closures were necessary.[9]

The NUM called a Special Conference to discuss the situation.

At the Special Conference Paynter argued that Lee's 1 July statement 'revealed completely new thinking as so far as the Government were concerned.' The principle underlying government policy was the primacy of 'consumer choice' which had characterised the Conservative government's policy, and coal was now facing 'unplanned contraction' and although the Central Coalfields of Yorkshire and the East Midlands would benefit other coalfields would be virtually wiped out.[10] Delegates united in condemnation of government policy. Jack Dunn (Kent/NEC) insisted 'Government policy is not only discredited, but it is a political betrayal of the British miners and their families' whilst Alf Hesler (NEC/Durham) warned 'we cannot and won't agree to acceleration, because any form of acceleration...would not give us sufficient time to make the necessary arrangements'. A young Yorkshire delegate, Arthur Scargill, condemned the NEC's policy as 'appear[ing] to be an acceptance of [closures] provided we can get it brought about with as little harm as possible to the people who are going to be affected'. Bill Rowe (NEC) concluded 'There is nobody more loyal to the Labour Party than I am, but I feel it is time that we have got to say very, very forthrightly, "So far and no further". We have [to] resist where we feel that we are going to be let down badly by the Government'.[11]

Fundamental to the NCB–Government relationship was the insistence that pit closures were not the responsibility of the Minister of Power but an operational decision for the Board. The Categories had been drawn up by the NCB on government instructions so in effect, any pit in the C Category would close on government instructions irrespective of the consultation procedures. Lee argued government policy was designed to help the NCB 'eliminate grossly uneconomic pits as quickly as possible', so while 'The Government had accordingly approved the closure programme submitted by the Board...the Minister insisted that he had no statutory powers under which he could impose obligations on the Board in respect of the actual timing of the closure of individual pits.' Lee insisted closure was still a matter for the NCB in conjunction with the NUM.[12] However, in 1968 Robens told a Select Committee:

> There was considerable pressure put upon the Board to accelerate the pit closure programme...there was an occasion when I am sure the Ministry felt we were not closing pits fast enough and when it was virtually demanded of me that I should submit a list of all the pits to be closed quarter by quarter.[13]

Ministers were infuriated by Robens' criticisms especially as some Board members, including his deputy, Derek Ezra, endorsed rapid restructuring. Robens was, however, politically un-sackable. If sacked he would go public, as he would if the minister issued a direction of which he disapproved. Nor, for the same reason, could he be allowed to resign. Despite wanting to get rid of him, Richard Marsh, Lee's successor as

Minister of Power, refused to accept Robens' resignation when the Aberfan disaster inquiry in August 1967 blamed the NCB for the deaths of 144, of whom 116 were children, when a pit tip buried a school on 21 October 1966.[14]

Aware of the disquiet this programme would cause in the coalfield constituencies Labour produced a pamphlet explaining and justifying the policy. No government, the pamphlet argued, could guarantee a 200m ton industry because 'Such a policy would not make economic sense. Nor would it make social sense.' The Government, the NCB and NUM had agreed to concentrate output on the pits with the best conditions and highest productivity, a policy obstructed by 150 high cost pits which:

> the industry has to carry. Their existence threatens the future of the majority of good and newly reconstructed pits. So long as the good pits have to subsidise those which do not pay their way, prices will be at a level where even the good pits find it difficult to sell their coal.[15]

Labour's 1966 manifesto, *Time For Decision*, declared that 'The best available estimate for the market in coal in 1970 is 170-180 million tons. We stress that this is an estimate and in no sense a limitation. Everything depends upon efficiency, costs and the resulting prices. If more can be profitably sold then no barrier will stand in the way of expansion.' The 1965 closure programme was a good example of the pressure that could be brought to bear on a formally independent organisation. This pressure was described by the NCB as 'very great and quite contrary to the underlying philosophy of Board–minister relations.' The absence of formal directions had 'not impeded successive Governments from exercising considerable pressure of the Board by more informal methods, which are not published.'[16]

After the 1966 General Election Lee was sacked by Wilson as Minister of Power and replaced by Richard Marsh. At his first meeting with the miners Marsh reiterated that protection for the coal industry would continue but raised a new factor in the fuel equation, the huge reserves of gas in the North Sea. The problem was that despite the support being given to coal, its markets continued to decline so 'one had to face up to the question of just how much would have to be paid to protect coal adequately.' He continued:

> If the existing level of protection would not persuade the consumers to buy, one could not compel them, and it was to be wondered just what degree of protection would persuade them. Uneconomic pits were averaging up the cost

of coal. If coal was to be sold, the price had to be brought down to a level where it was more competitive.[17]

Reflecting on its failure to deflect policy the NEC acknowledged government was trying to 'restore the industry to a healthy and competitive position' but its policies 'will reduce the industry to a level well below that forecast for 1970.'[18] Ford expressed the frustration and incomprehension current in the NUM when he confessed at the 1966 Conference that 'today, instead of the kind of security and lasting prosperity that we all imagined would be a feature of the coalmining industry under State control and ownership, we have an industry riddled with frustration, cynicism and doubt as to its future.' The only feasible response was to continue to work for government intervention and secure the phasing of the contraction.[19] Composite III dealt with closures and was interesting because it hinted at the end of the social contract at the basis of public ownership which had been intended to produce coal economically and efficiently but meant that profitable pits and coalfields would support the unprofitable. However, according to Bill Carr, a Yorkshire delegate, 'This social ideal has now been scrapped. Pits are being closed, whether we like it or not, irrespective of the long-term future...a grim reminder of the less fortunate days when coalowners were belaboured for this reckless dismantling of our industry'.[20] Part of the social contract involved the eschewing of industrial action by the NUM, pit closures jeopardised this self-denial.

In December 1966 Marsh wrote to the NEC informing them that he was reviewing the 1965 fuel policy. This review was prompted by the Ministry of Power's use of computer driven econometric models which had been introduced to improve the rationality of decision making and reduce political controversy. The NUM identified three likely consequences: production would be speedily concentrated on the most modern and profitable pits; accelerated mechanisation would force down labour costs; and maximum machine utilisation would radically change working patterns.[21] An informal meeting with Marsh took place on 12 January 1967 to discuss the widespread rumours of further closures. Although the policy review was far from complete Marsh hinted that further closures could not be ruled out but that the government would provide more finance for the social costs of closures. The NEC's Economic Sub-Committee clung to the view that 'the industry's size will still be determined by its ability to compete' and it pointed out that the since nationalisation the NUM had accepted 'the need for re-organisation and a large scale programme of concentration.' However, the exploitation of natural gas from the North Sea, the decision to build AGR nuclear reactors and the expansion of domestic oil refining had increased immeasurably the complexity of the NUM's environment.[22]

The 1967 Fuel Policy Crisis

Throughout the spring of 1967 rumours circulated in the pits that the industry would be cut back to 140 million tons by 1970. Marsh denied these rumours and reiterated that the NUM would be consulted fully, a statement which was greeted sceptically as the 'disregard by the Government of the views of the Union stemmed from an assumption that the NUM would co-operate in any measures the Government proposed.' Since 1965 the NUM's stance had been based on 'the premise that the industry's productive capacity would be maintained at around 170 million tons per annum' and concluded 'the Government should be warned of the consequences of a further run down of the industry.' One month later these rumours were extended to include the reduction of the industry to 80 million tons by 1980. Rumours were fuelled by the scenarios being run by the Ministry of Power and the discussions between ministers, civil servants and public sector fuel industry chairmen at which the NCB were known to be seeking increased government help to meet the social costs of a further rundown.[23]

When Marsh met the NEC he acknowledged that for the past fourteen months the ministry had been exploring various scenarios, a meeting of nationalised fuel industry chairmen had been called for 18-20 May 1967 at the Selsdon Park Hotel to discuss them and once a policy was determined it would be put to the Cabinet for discussion. No final decisions had been taken but coal was too expensive and the ministry searching for 'the price level at which consumer behaviour would change' and he would make a statement when the Coal Borrowings Order came before Parliament in July. The Selsdon Park conference served to confirm the shift of energy policy to nuclear and gas and revealed the power of the nuclear bias in energy policy. Its conclusions were leaked.[24]

The NEC proposed an Emergency Resolution to replace Resolution 41 at the forthcoming NUM Conference which called for the maximum use of indigenous coal and that North Sea gas and oil be used to replace imported oil.[25] In the House of Commons the NUM's sponsored MPs warned Marsh and the Cabinet of the dangers of their policy. Shinwell asked if Marsh was 'aware that the mining community is almost in revolt against the Government'? Edwin Wainwright, the ultra-orthodox MP for Dearne

Valley in the Yorkshire coalfield, complained that ministers gave 'the impression that they are indifferent about the future welfare of the men in the industry'; and Ness Edwards regretted that 'Promises have been given and those promises have not yet been carried out.'[26]

The NUM knew what was coming. Even Ford evinced a deep sense of betrayal and foreboding because:

> it has been our unfortunate experience in the last year or so, that speculation which we discounted at the time because of our faith in policies and attitudes to which we believed the Labour Movement was firmly committed, were later found to be well-founded in that they conformed fairly accurately to subsequent Government pronouncements.[27]

On the Wednesday of the NUM Conference Marsh delivered a long justification of Labour's policy, contrasting this with the policy of the previous Conservative government and justifying the need to modernise the coal industry as part of the government's wider programme of modernisation. He also urged delegates to remember the scale of the economic crisis facing the government. The November devaluation crisis was about to explode and there was uncertainty over future oil supplies and their cost because of the Arab–Israeli War in June which seemed to confirm the NUM's warnings of the danger of relying on imported oil from unstable regions.[28] Marsh's speech was intended to influence the Thursday debate on fuel policy which was monitored by Marsh's Parliamentary Secretary and the head of the ministry's Coal Division.

Paynter was deeply critical of the government ('we are having to face harsh experiences these days, and we are entitled to use hard words') and he blamed government policy for undermining the mineworkers' faith in Labour which had been reflected in the municipal elections.[29] The debate on the NEC's Emergency Resolution was of little real significance and it passed *nem con* but the 1967 Conference was pervaded by a sense of brooding anticipation. Bill Rowe expressed this clearly; 'I want this government to recognise now that the social and political costs of [this] policy could be disastrous to a Labour Government. The red light is on! Our people at the last County and Municipal elections, to be rather rude, voted with their backsides.'[30] Delegate after delegate expressed their lack of confidence in both Marsh and the government. From the left Jack Dunn (NEC/Kent) stressed the proven habit of Conservative and Labour ministers of denying any intention of cutting the industry and then doing exactly that. From the right Joe Gormley (NEC/North Western Area) admitted that he had always believed Labour would never 'pay lip service to the oil barons, to the detriment of those people, who, over the years, have been their strongest supporters...I also did not join the Labour Party to support a policy that created unemployment...I have been so embarrassed in these last few months I have thought of resigning from the [NEC] of the Party.' Speaking at his first Annual Conference Arthur Scargill declared 'I never heard flannel like we got from the Minister' who 'said that we have nuclear power with us, whether we like it or whether we don't. I suggest to this Conference that we have coalmines with us, and we still have, but they did something about this problem: they closed them down.'[31] That coal and the mineworkers had been betrayed by Labour was accepted without dissent.

Marsh was to make his Commons statement on 18 July. On 13 July the National Officials met Marsh, and on 15 July they met Wilson, Callaghan (Chancellor of the Exchequer) and Ray Gunter (Minister of Labour). Also present were Robens and Bill Webber, the NCB's Industrial Relations member. Ministers' attention was drawn to the social costs of the government's policy, a request for a subsidy to keep pits open until alternative employment was available was made, and the NUM objected vehemently to the CEGB's proposal to build a nuclear power station at Seaton Carew on the edge of the Durham coalfield.[32] The National Officials came away from the meeting convinced that Wilson had been favourably impressed by their arguments, so Marsh's statement came as something of a shock. Crossman had 'specially arranged' an all night sitting on the Coal Industry (Borrowing Powers) Order to allow the Miners' MPs to complain at length about government policy 'before accepting their fate.'[33] Econometric modelling had indicated a 140 million ton industry in 1970 but Marsh had rejected this on social and economic grounds as too low in favour of consumption of about 155 million tons achieved by subsidising the CEGB's coal purchases. The new variables in the equation were the AGRs and North Sea gas which would displace large amounts of coal, so further contraction was inevitable.[34] Normally docile Miners' MPs queued up to attack Marsh and the government and Crossman described the statement as 'devastating' but he believed the government would get away with it as:

> it was clear that provided they could make their protest these miners felt they were bound to support the Government in an action which really meant the destruction of the mining industry. What these Miners' MPs showed was a not very edifying loyalty, because people should not be as loyal as that to a Government which is causing the total ruin of their industry...I was shocked by their pathetic lack of fight.[35]

Edwin Wainwright warned Marsh that 'he must not take it for granted the section of mining constituency Members will always allow him an easy time...we are sometimes a little doubtful about his conduct in regard to the mining fraternity. This will depend on what happens in the next few months. We shall watch him carefully.'[36]

Wilson, Callaghan and Michael Stewart (the Foreign Secretary) were the principle guests and speakers at the 1967 Durham Miners' Gala where they addressed 'a solemn, unenthusiastic audience.' Afterwards they met Robens, Paynter, Ford and members of the NCB and NEC where 'they were told the real facts about the disaster the Government's closure policy was precipitating.' Ministers were worried at the wider effect of another 150,000 unemployed at a time when Labour was doing badly in the polls, and Wilson and Callaghan had been shaken by their hostile reception at the Gala. Wilson told colleagues he was considering postponing the closures, despite Marsh's opposition, in order to reduce the number of unemployed. A major crisis was developing up but no one in the government, not even Marsh, knew exactly which pits the NCB proposed to close and the Ministry had not analysed the social consequences. Crossman was aghast because 'the lesson of this meeting was the appalling fact that after three years the Labour Government had evolved neither an instrument for assessing the social impact of its actions nor an instrument for ameliorating that impact upon the community.'[37]

Ministers were under great pressure even before the publication of the revised fuel policy White Paper and, according to Robens, Wilson was ready to compromise whereas Marsh was prepared to tough it out. Robens warned Wilson that the miners 'were at breaking point over pit closures' and only their residual loyalty to Labour and the NCB's mitigation of closures 'had so far prevented industrial disputes.'[38] There was actually little evidence of active unrest in the pits and this claim is an example of Robens applying pressure on a vulnerable government. Another favourite Robens tactic was, according to Marsh, to schedule pit closures to coincide with sensitive political events such as by-elections. He was then forced into the embarrassing situation of having to try and persuade Robens to postpone pit closures.[39]

On the eve of the 1967 Labour Party Conference at Scarborough, NEC members met Wilson, Peter Shore (Minister for Economic Affairs) and Marsh to discuss the crisis in the industry. To defuse the NUM's anger and help smooth what promised to be a rough conference Wilson revealed that, in response to 'certain difficulties', the NCB had agreed to postpone all currently projected closures until after the end of the year. Wilson emphasised postponement 'did not mean that there had been any changes in the Government's fuel policy objectives, or on the future of the coal industry as outlined by the Minister of Power.' Somewhat patronisingly, Wilson also offered to allow one security cleared NEC member to spend a week at the Ministry of Power to see how the fuel use projections were formulated. NEC members were appalled. 'These people', the minutes recorded, 'expected something better from a Labour Government' and the NEC made 'it clear that they were far from satisfied.'[40] On the same day that mineworkers protested in London against the policy, the NEC's Economic Sub-Committee met Marsh in a final attempt to secure more support for the transition costs and a commitment to build new coal-fired power stations (the coal-nuclear comparative cost analysis was not yet finished). Marsh gave no ground.[41]

The 1967 White Paper amplified that of 1965 and took into account North Sea gas and the AGR programme. So:

> Further decline in the markets for coal could not be prevented even by holding back the expansion of nuclear power and the development of natural gas unless the present level of coal production were raised to an extent which would lead to a big increase in the general level of energy prices, or unless coal were heavily subsidised. But excessive protection for coal would lead to a misallocation of manpower and capital to the detriment of the economy as a whole.[42]

Cost had been the main driver of coal policy since the mid-1950s and government policy was to manage coal's decline sympathetically. The industry's decline was 'not the result of government policy: it reflects a continuing trend in consumer preference' and 'the modernisation of the coal industry and its concentration on the most economic coalfields and collieries must go forward. Only in this way can the coal industry remain viable.'[43] Viability depended on increased productivity but mechanisation would supplant labour so continuing job losses were inevitable. In the absence of a general subsidy government would fund the social costs of transition and modernisation where these could not be met from the industry's own resources. The White Paper's object was 'that the industry's contraction during the next few years should be brought about with the least possible hardship in mining communities and without interruption of the modernisation of the industry.'[44] The rundown programme assumed a manpower loss of 35,000 per year up to 1970-71. The Most Desirable Policy scenario balanced the social and political costs of closures with the scenarios explored by the Ministry of Power's econometric modelling. These produced the April 1967 Estimates which were at the heart of the White Paper and which were discussed extensively at the Selsdon Park Conference in May 1967. The 1975 level was achievable but the 1970 figure pointed to too rapid a contraction in the medium term, hence the decision to hold coal demand at 155 million tons.[45] To ease the transition the CEGB would be subsidised to enable it to burn more coal between 1968 and 1971 than it otherwise would, and a crucial assumption of the transition policy was that redundant mineworkers would be speedily absorbed into the labour market and that the mining communities would rapidly adapt. Coal policy was 'a quest not for the optimum level of coal output, but for an optimum rate of contraction.'[46] The philosophy underpinning the White Paper was that:

> market forces, with even the present excessive degree of preference for coal, will force coal output down over the next ten years pretty drastically. We see no reason to interfere with this process, save in the short run, when the pace of contraction, compared with previous expectations, might be too sharp to be accepted psychologically or absorbed economically, hence policy consists simultaneously proclaiming the inevitability of long-term contraction while striving to limit it in the immediate future.[47]

This was further complicated by the NCB's financial regime. Since 1961 it had been permissible for some collieries to make losses even when the industry as a whole was meeting its financial obligations but, given the current financial state of an industry where six out of eight NCB Divisions were loss making, government thought it prudent to write off 'unremunerative investment'. Capital investment continued and £455 million was budgeted down to 1971 and after allowing for the lower level of coal demand the NCB's borrowing limit was increased to £750 million. From 1961 the NCB was expected to balance their accounts year on year over five years after interest and depreciation, it was relieved of this obligation in April 1965 until the financial year 1967/68. The Labour government imposed new objectives for the nationalised industries insisting they:

> increase the productivity of both labour and capital employed; to raise the rate of new capital formation; to ensure that new equipment is as technologically advanced as possible and is effectively deployed; to increase the profitability of new investment; and to obtain the maximum return in terms of the production of goods and services.[48]

Pricing policies should avoid cross-subsidisation (integral to the original purpose of public ownership) and there should be continued downward pressure by management on wage costs to maximise efficiency.

By November 1967 Marsh and Robens were in open conflict, the Miners' MPs were on the verge of mutiny and Marsh was 'beginning to realise that the extraordinary rapid and ruthless closure of the pits which he's forced though Whitehall and Cabinet is going to be pretty unacceptable outside.' Crossman continued, 'Dick Kelley has already denounced the Minister of Fuel, and our friend Joe Gormley...has stated that he's prepared to start a new miners' Party because the Labour Party is betraying them.'[49] Despite 'the very ugly parliamentary situation' Crossman announced that the government's intention was to seek parliamentary approval for the White Paper but 'found that the Miners' Group were raging round saying that they bloody well weren't going to approve.' Shinwell threatened the government with mayhem and Crossman told the Cabinet he feared 60 to 70 PLP members might refuse to vote for the government's motion.[50] Wilson and Peter Shore wanted to phase the closures over 18 months, Marsh and Robens were opposed because coal was piling up at the pitheads, so as an interim response the Cabinet agreed to withdraw its parliamentary motion. The White Paper was a provocation too far; 'Under the Tories, closures were going on nearly as fast as under us. The only difference is they went on quietly without fuss whereas we are making an enormous bother about them and also giving enormous concessions...and getting nothing but abuse for doing so. What a mess.'[51] On 23 November the Cabinet withdrew the White Paper in order to reconsider the implications of the devaluation of Sterling on 20 November for the government's energy policy.

The withdrawal of the White Paper, the NEC noted, 'did not mean that one could expect a significant improvement in the prospects for coal.' Not surprisingly the NEC was asked to consider requests from various Areas for 'further action' and in response the NEC called a Special Conference to discuss tactics and sought a further meeting with Wilson. Between January and March 1968 21 pits were scheduled to close and the fate of four more had yet to be decided so 1968 would be the worst year for closures yet.[52] A battered Marsh agreed with the NUM that the government would shoulder the costs of social dislocation but 'The real problem was how far it was possible to distort market trends. In spite of the measures already taken to distort the market in favour of coal, the decline in coal consumption had continued.' Nevertheless the assumptions underpinning the White Paper were being 're-assessed' in the light of devaluation of Sterling but the

Economic Sub-Committee doubted there would be a change in a policy which had now been pursued for a decade.[53] The crisis in the industry was blamed on the government which was accused of pursuing the free-market policy of their predecessors. To the annoyance of the NUM Wilson went over the same ground as at the Scarborough meeting, reiterating it was not the government's intention to run down the industry and denying policy was market driven.[54] At Selsdon Park in May 1967 Robens had persuaded Marsh and the Ministry's civil servants to increase projected coal demand and Robens continued to press in Whitehall for an independent comparative analysis of coal and nuclear generating costs. Robens reminded CINC that the 16 postponed closures would have to proceed as soon as possible in 1968. In September 1967 NCB manpower (392,900) was projected to fall by 117,900 to 275,000 by 31 March 1971 and by a further 115,000, to 160,000 by the end of 1975. Between 1967 and 1975 manpower would fall by 232,900 (59.3 per cent) but these estimates assumed an output per manshift of 50 cwts by 1971 and 75 cwts by 1975. Robens called for a joint NUM–NCB campaign to boost efficiency and reduce costs, the postponement of any new nuclear power stations and the building of three or four new coal-fired power stations.[55]

The recognition that there was no possibility of changing the fundamentals of energy policy by conventional lobbying led some in the NUM to argue once again that 'some form of industrial action would have to be resorted to' such as an overtime ban, a one-day stoppage in an Area over a particular closure, or even national action. There was, however, widespread doubt that the membership would endorse industrial action and opponents argued forcefully that 'industrial action would do no good for the industry or the men employed in it [who] would loose far more than they could possibly gain.' A Special Conference held in March 1968 was critical to the political development of the NUM because:

> the NEC should clear up whether or not strike action was in the minds of the members and establish whether or not it was a weapon the committee could use in an attempt to change the existing climate. The alternative, it was argued, was to accept the inevitability of contraction and to concentrate our energies with the Board to improve efficiency in order to make coal more competitive and thus increase the coal burn.

This statement encapsulated the political strategy of the NUM since the General Strike. A policy change on industrial action would have repudiated the Union's post-1926 ethos and the conviction that nationalisation represented a shift in power relations in the industry. This was literally unthinkable to a majority of the NEC and 'a proposal that there should be some kind of industrial action was rejected.'[56]

Those opposed to industrial action argued it would have no effect. Tommy Burke (Yorkshire/NEC) argued the mineworkers lacked any economic clout so a national strike would damage the industry, neither the government nor NCB would pay any attention to a one-day strike and guerrilla strikes would fragment the workforce. Burke reminded delegates that 'nationally we have never had any industrial action since 1926. Just throw your minds back. I can remember it quite well, but a lot in this audience cannot.' Conditions had changed 'and the type of man you are dealing with today is totally different from the type of man you had at that particular time' and Burke articulated a straightforward embourgeoisement thesis; 'Today our lads have set on things like £800 cars, £3,000 bungalows, all that kind of thing and it all has its effect. It is a different system of life, and I do not think for a moment that they would jeopardise all those things for the sake of a strike which, in my opinion, could have no impact whatsoever'.[57] Arthur Pratt, another NEC member, agreed the NUM was at a cross-roads: 'Strike or Talk?' He favoured talking; 'look at the vacant seats. We should have filled this damned place at one time. We are insignificant now, and if we had a strike tomorrow (1) would we carry our members with us? I can speak for those in the Midlands. We would not. And (2) would we stop a plane going out from Heathrow Airport today by stoppage? We would not. Would we stop a train going out of Euston? We would not.' Alf Hesler, also an NEC member, endorsed this and reminded delegates they would be striking not just against the government but against a Labour government 'that we put in after experiencing for 13 years the tyranny and exploitation of a Tory Government...I still think the Labour Party with all its failings is a far better choice than a Tory Government'.[58] Looming over the debate was the spectre of 1926:

> We had right on our side in 1926, and we still could not win. I can remember the national leaders of that time coming round my local township, "Tighten your belts! We shall win. Not a penny off the pay, not a minute on the day." ...we went back with a few pennies off the day, and a lot of minutes on the day ...if we could not win then, how much more chance do we stand of winning now, which must surely go down as the weakest moment of our industrial strength.

Dick Main (NEC) warned 'we have had the barometers out...and we believe that if a ballot was taken...it would record a vote against strike action'.[59] A Scottish delegate believed 'there has been far too much talking and it is about time we had some action' and he noted that after 1947 the NUM had never used its 'industrial strength to defend the interests of the miners', hence their current situation. Apathy in the coalfields reflected a failure of national and local leaders who had been 'elected to fight for and defend this industry and the membership, not to preside over the destruction of large sections of it.'[60] A 64-year-old unidentified delegate who had worked in the pits for 51 years and who had experienced 1926 described how his 'monthly branch meeting last Sunday...discussed this matter, and we thought we had reached the end of the road...we took a decision that we should have some sort of industrial action.'[61] In a riposte to Pratt, Jock Kane, a full-time official from Yorkshire, argued the seats were 'not empty as a result of industrial action. They are empty because we have failed to influence in any way this Government or the previous Government to halt the run-down of our industry.' Kane welcomed the palliatives but Labour's abandonment of Socialism meant nothing could be achieved by parliamentary means so 'it is necessary now to introduce some form of action that will sharply bring to the notice of the powers that be the fact that the miners are no longer prepared to go along the road they have of continued closures, redundancy and destruction'.[62] The lobbying strategy was approved without a card vote.

After the Special Conference the steam went out of the closure issue. In his last speech to an NUM Conference before his retirement Bill Paynter railed against the 'outlook of "Close the bloody lot"' in the coalfields, blaming the government, rather than the NUM's leaders and its members, for the demoralisation in the coalfields:

> I know I sound bitter, and I am bitter. Those of you who went on the trip yesterday went over the mountain and down into the top of the Rhondda Valley...there are 130,000 to 140,000 people concentrated in those narrow valleys with nothing else there but the pits. And there are only three pits now working in the Rhondda Valley...[63]

Paynter was correct, there was much to fight for. In 1967 the NCB was still the largest employer in Western Europe with a wage bill of £8 million per week, an annual turnover of £800 million and annual capital expenditure of £80 million. It employed 400,000 wage earners and provided 55.4 per cent of the country's total inland energy needs. The NUM did win some concessions. Between 1959 and 1970 coal imports were banned and from 1961 a duty of 0.83p per gallon was levied on heavy fuel oil (increased to

1.01p in 1968) and the Coal Industry Act (1967) subsidised the CEGB burning 6 million tons of coal a year more than it would have used on purely commercial criteria. Crumbs of comfort were offered by the Economist Intelligence Unit's report *Britain's Energy Supply* (September 1968) which predicted a rate of decline slower than that envisaged by the 1967 White Paper and at the Sunningdale Conference in November 1968. Whilst Sunningdale was not designed to help coal, Mason used it to articulate his belief that closures were taking place too rapidly but 'the policy', Robens noted, 'remained unchanged'. Derek Ezra, the NCB's head of marketing, predicted a fuel gap in early 1970s which might have to be filled with imported oil because of the rundown and Robens felt 'that the [CINCC] could be optimistic – cautiously optimistic.'[64]

Adding Insult to Injury

In January 1964 the NEC noted that increased difficulties with the NCB over pay was leading to outbreaks of unofficial industrial action. The NCB's 'attitude could lead to many difficulties for the Union leaders and there could also be a tendency for fragmentation within the Union with consequent difficulties for the Board and the Union.' The Board accepted there was growing discontent in the industry over pay but insisted it was under no pressure from ministers. Indeed, Robens stated that if there was such pressure he would demand written instructions. The NEC concluded that it 'had to accept that this was the best that could be obtained in the circumstances' and agreed to ask the Special Conference to approve further talks.[65] The Special Conference called to discuss pay was extremely bad tempered and there were frequent warnings that the NUM was standing at a crossroads. Herbert Parkin (Derbyshire) argued the NUM's negotiating position would be strengthened by an overtime ban, a suggestion which horrified the majority on the NEC. Despite their anger the perceived absence of a viable alternative led delegates to agree to allow its negotiators to meet the NCB again.[66]

In the month of talks there was frantic lobbying throughout the NUM to secure support for industrial action. So successful was this that Ford felt compelled to warn the Special Conference that if some delegations had come with a prior mandate they were in breach of NUM rules and conventions and he threatened to exclude them from the Special Conference. Paynter pointed out that the NUM's wage increases since 1958 compared favourably to those achieved in other industries but conceded that the main issue before the Special Conference was the NUM's reaction to the NCB's intransigence despite the marginal improvement in its offer. Paynter identified four possible responses: first, the tendering of 14 days notice followed by a strike, but was the difference between the NCB's offer of 47.5p and the NUM's claim of 75p worth jeopardising 'the security and employment of thousands of men' as the NUM had no allies and would be confronting the NCB *and* government? Also a general election was in the offing and a major strike in the coal industry might damage Labour's electoral prospects. Second, a one-day strike would be a token gesture and 'is [not] going to force the NCB to its knees' especially as coal demand was falling as summer approached. Third, a work-to-rule was rejected as this could lead to mineworkers being prosecuted for breaches of the Mines Regulation Acts; and fourth, an overtime ban would breach the Five Day Week Agreement, the NCB would instantly break off negotiations and this would damage the men's interests more than it would the NCB. Paynter concluded, 'there is no effective alternative before us in this situation.'[67] Scotland, Wales and Derbyshire opposed this recommendation but most seriously for the NEC, the large Yorkshire and Nottinghamshire delegations were either split or undecided on their attitude to the NEC's recommendation. Faced by the possibility of industrial action the NEC raised the 1926 spectre:

> Some people say it was a moral victory. It was a material disaster for the following generation after 1926. In 1963 the last debt to the miners in the Northumberland Area was written off...But the biggest disaster was the generation that was affected by virtue of the strike – the children of that generation.[68]

After a brief adjournment the NEC's recommendation to accept the NCB's offer was put and defeated on a show of hands. Ford's suggestion that the matter be put to the membership in a pit-head ballot was accepted with alacrity. To some delegates this seemed to indicate political change on the NEC but the right-wing majority's motivation was simple: 'there has', Jack Robinson (Durham) declared, 'been quite a lot of so-called evidence of so much unrest' so a national pithead ballot would provide a definitive and irrefutable measure of militancy in the NUM.[69]

In the ballot 73.1 per cent voted in favour of the NEC's recommendation to accept and 26.8 per cent voted against this, and of the NUM's constituents only Scotland recorded a majority (59.8 per cent) against the NEC's recommendation. Even if Scotland's vote had been replicated nationally it would have been insufficient to authorise strike action in a Rule 43 ballot which required with a two-thirds majority to authorise industrial action. Four coalfield Areas (South Wales, Kent, North Western, and Yorkshire) recorded a vote against the NEC's recommendation which was above the national average vote against (71.1 per cent in favour, 28.9 per cent against) for coalfield sections of the NUM. The percentages voting against the NCB's offer ranged from Scotland's 59.8 per cent (highest) to South Derbyshire's 16.3 per cent (lowest); those above the national coalfield average vote against were on the left (or as in the case of Yorkshire) contained an increasingly powerful left-caucus and those below the average were on the right of NUM politics. In the non-coalfield NUM Sections only Group No.2 produced a vote against above the coalfield average (33.5 per cent) and in the Sections the average vote against was 24.7 per cent compared to 75.2 per cent in favour. The bulk of the NUM Groups were on the right of NUM politics.[70] The conviction that there was little stomach for industrial militancy in the NUM was confirmed by the 1964 ballot but nearly one-third of coalfield membership and one-fifth of the functional area membership (a combined percentage of 29.6 per cent) were prepared to vote against the NEC's recommendation to accept the NCB's offer. Those Areas polling at or above 30 per cent against formed the core of the left faction that was to take control of the NUM at the end of the 1960s. The key swing Areas were Durham, a traditionally a right-wing bastion which was suffering heavily from pit closures and whose right-wing orthodoxy eroded after Sam Watson's retirement in 1963, and the huge Yorkshire Area which contained a significant left caucus in, but not confined to, the Doncaster pits.

That collective bargaining in conditions of full employment caused inflation was a touchstone of post-war economic management. Despite the difficulties of the Attlee government between 1948-1950 and the failure of Macmillan's 'three wise men', the 1961 'guiding light' and the National Incomes Commission, the preferred official solution was a prices and incomes policy. Whilst in opposition Labour and the TUC moved towards agreeing voluntary pay restraint as part of a package including price controls and economic growth. Before coming into office neither Labour politicians nor trade union leaders envisaged detailed intervention in collective bargaining and no joint economic strategy was worked out before October 1964.[71]

Labour had been elected in 1964 committed to 'the planned growth of incomes', essentially a voluntary incomes policy. Miners' wages had been a problem for all post-war governments and in April 1964 the NUM had held its first national ballot on a wage offer. Pressure on wages was a major reason why the NUM was so keen to press ahead with the negotiation of the National Power Loading Agreement (NPLA).[72] The 1964 NUM Conference had demanded substantial increases for the lowest paid as well as for a general increase in national rates. Negotiations had begun with the NCB but the October General Election and the development of Labour's 'planned growth of incomes' policy led to their suspension. Confronting a major economic crisis, the Labour government sought an agreement on wages with the TUC which agreed in December 1965 to voluntary restraint. George Brown at the Department of Economic Affairs (DEA) negotiated an agreement with the unions and employers whose 'basic principles...will govern future incomes policy. I am not suggesting...that this is a factor which ought to determine the wage claim we make, but it obviously is a factor that has to be taken into account when we are examining the general political and economic climate'. The NEC were willing to trade wages for a fuel policy but there was a strong undercurrent of feeling to the effect that if there was no fuel policy the NUM should not accept wage restraint.[73]

The NUM was committed to supporting the TUC policy and supporting the Labour government. The NEC agreed to attend the TUC Conference of Executives (30 April 1965) called to discuss the document *Productivity, Prices and Incomes* which formed the basis of the accord with government, with a predisposition to support it but at the same NEC meeting it was decided to submit a claim for a substantial increase for daywagemen. The NEC accepted the planned growth of incomes policy but argued that the time was not yet ripe for such a policy. Wage settlements in mining were above the 3.5 per cent norm suggested by the TUC and the NEC feared 'difficulties with their members' if 3.5 per cent was the limit. The NEC also feared pay policy would disrupt negotiations with the NCB who could cite government policy as justification for resisting the NUM's claim. So the NUM rejected the norm but accepted the principle of the TUC's policy and in defiance of the TUC proposed to submit a wage claim in excess of 3.5 per cent.[74] The NUM's negotiators warned the Board of 'the strong dissatisfaction with the inadequacy of the present wage rates and there was not much room for compromise...This mood existed throughout the industry and could affect its whole future.' The NEC accepted the need for an incomes policy but concluded 'that the time was inopportune' because the current industry going rate was above the proposed 3.5 per cent limit and the cost of living risen. It would therefore be difficult for the NEC to accept the TUC's recommendation as the NUM had already submitted a claim and could not ignore a policy sanctioned by its Conference. However, the NEC agreed to support the General Council.[75] The NCB's offer was within government policy and was rejected by the NEC on the grounds that it would not increase the wages of the low paid nor retain skilled labour. The NCB's 'dilemma was to balance this view with the job of keeping pits open', so wage increases would have to be financed from productivity gains. Furthermore 'The Board's objectives were governed by statute *and the pressure from the present government was just as strong as from the last government*' and in consequence 'they were in no bargaining mood and they emphasised that in the interests of all the men in the industry, they could not take the offer any higher.'[76] The NCB made some minor concessions and the NEC accepted the offer reluctantly as the best obtainable in current circumstances.

In July 1965 Labour suffered its first major economic crisis. This led to the beefing up of the National Board for Prices and Incomes (NBPI) and the TUC reluctantly agreed to vet pay claims in the knowledge that the government was prepared to institute a statutory pay policy if the TUC failed to deliver voluntary restraint. The 1965 NUM Conference was dominated by the move towards NPLA and the accelerated pit closure programme and on wages Conference accepted NEC advice to remit all wage resolutions. In the Autumn the NEC submitted its second wage claim partly to respond to pressure from the low paid and partly to divert attention away from its difficulties over pit closures. This claim was counter to both the letter and spirit of the joint government-TUC policy and the National Officials met the TUC's vetting panel to discuss the claim. They insisted the claim was in line with George Brown's statement of 30 April which permitted claims outside the 3.5 per cent norm if such a claim was in the national interest and was needed to retain skilled manpower. The NUM had been awarded 5 per cent in June and was now seeking a further 5 to 6 per cent which, the TUC panel argued was bound to be cited by other unions. The TUC suggested 'a sacrifice on the part of the miners' and the postponement of the second claim in the wider interests of the Labour government and movement. This was rejected by the NEC.[77] In February the NEC expressed 'bitter resentment' at the NCB's rejection of its claim and unprecedentedly resolved that 'a ban on overtime working in the coal mining industry be imposed' from March.[78] The decision to call a ban in one month's time was falling when coal demand was largely symbolic but the Special Conference (18 February) was significant because it debated the NUM's position vis-à-vis the Labour government.

The NUM was operating in a very complex political environment and on wages was under pressure from the NBPI, the TUC vetting panel, other unions and its own members. The NCB was facing serious financial problems and for both it and the NUM the restructuring of wages took primacy. Yet the NEC agreed to an overtime ban acknowledging that:

> We have to recognise that, and it is, I suppose, in the nature of things and especially in relation to our history as a Miners' organisation since 1926, rather a drastic and dramatic action that the [NEC] have decided to follow in these circumstances. But we feel that we are justified in taking this decision.

Ford strongly opposed the ban and warned it might cost jobs but 'Unless there is some softening of the Board's attitude...this ban will be implemented.'[79] For the right in the NUM the overtime ban was a negation of all they had advocated since nationalisation and so radical a departure did it seem, even those sympathetic to the ban were worried that this was a stunt. This provoked an interesting comment from Ford to the effect that 'it has been the [NEC] who have been accused of dragging their feet behind the membership. I am rather surprised this morning that we are now being accused of going ahead perhaps beyond where the membership would want us to take them.'[80]

Neither the NCB nor the Labour government wished for a confrontation with the NUM. Wage negotiations were becoming enmeshed with the all-important NPLA as when the NCB warned the NUM it might have to choose between NPLA or a wage rise. The NEC backed away from the overtime ban arguing that as the NCB was now taking the NUM's claim seriously there was no justification for the overtime ban which was cancelled on 26 March.[81] The Prices and Incomes Bill was introduced in the same week as the NUM Conference and the start of the Seaman's Strike (July 1966). Ford was well aware of the political sensitivity of the NUM Conference, Harold Wilson was due to attend and address the delegates but withdrew because of the economic crisis. Ford argued any Labour government was better than a Conservative government advising that:

> we must not ask too much of it too quickly, after all, Rome was not built in a day; and neither, let me remind you, has Utopia yet emerged in the Soviet Union after nearly half a century of dictatorship. We must not assume that, governing in the interests of the whole country, the Government's decisions will

always please us. We must not forget that fundamental loyalty to the Labour Party is needed as much now as ever it was.[82]

An incomes policy was needed as part of the government's response to the economic damage of thirteen years of Conservative government and Ford denied it would result in a wage freeze. The unions had helped put Labour in office and they had an obligation to keep Labour there. Critics of an incomes policy argued that the unions would be loyal to Labour if Labour was loyal to them. An incomes policy was seen by them as a major threat to union freedom and Labour was asking for concessions that the unions would never dream of giving to a Conservative government. A prices and incomes policy was only feasible under socialism and if Labour insisted on such a policy it would eventually destroy the government.[83] Powerful voices were raised in support of Labour's policy. This support took the form of arguments such as, 'What ever may be the shortcomings of the Labour government, who may make mistakes, they are more preferable than a Tory Government'.[84] Despite these pleas Conference voted by 243,000 to 241,000 to oppose government pay policy.

The Seamen's strike, the Sterling crisis and Wilson's refusal to devalue the pound, led to the 20 July measures including a six-month wage freeze followed by a further six months of severe restraint coupled with a twelve-month price freeze. The NBPI was also given compulsory powers. The TUC acquiesced and in April had agreed to revive its vetting role but the 1967 Congress voted to oppose any further statutory wage controls. In September 1967 the NEC discussed the wage freeze and the Prices and Incomes Bill, both of which had been issued after the NUM Conference. Both had aroused considerable hostility and demands for non-compliance by the NUM but others on the NEC noted that the NUM had supported government policy since 1964 and was obliged to continue to do so. The NUM Conference opposed both compulsion and the early-warning of claims. The NUM would have preferred a voluntary policy but as this was no longer on offer it was proposed that the NUM should declare its support for the TUC's acquiescence in government policy. This was opposed by the left who argued union policy had been determined by the 1966 Conference and so the NEC had no option other than to oppose the government and TUC policy. As President, Ford had the power to interpret the NUM's rules and he focused on the NEC's responsibility for managing the NUM's affairs between conferences. This he took to mean that in the case of unforeseen events, or when events had negated a Conference decision, the NEC could effectively ignore Conference. Ford ruled that the 1966 Conference decision did not preclude the NEC independently coming to a view on Wilson's 20 July statement and subsequent developments, and this interpretation was endorsed by a majority of the NEC. Embarrassingly the NUM had already submitted a resolution to the TUC based on the 1966 Conference resolution and this opened the NUM to ridicule as it would be committed to opposing compulsion and early warning but accepting the wage freeze which was seen as part of a 'trend of gradually introducing legislation restricting trade unions.'[85]

Pit closures and pay restraint had created opposition to government and NUM policy in the union and as an organisation the NUM was undoubtedly in a difficult and contradictory position. This was not lost on the NUM's leadership. Few miners read the NUM's 1966 General Election manifesto but it is interesting because it explained the union's position with remarkable clarity. 'You may well ask', the NEC wrote, 'how we reconcile this advice [to vote Labour] with our attitude over recent months when we have felt impelled to criticise the present Government on various aspects of their fuel and power policies.' Reconciling these two positions was achieved by the fact that Labour governments had done much for the mineworkers and (as happened between 1945-51) the mineworkers had accepted that their sectional demands had to be reconciled with the national interest by a Labour government which 'must of necessity...have regard to its total responsibilities.' Labour was a national government embracing the interests of the miners as part of the wider community and, in any event, the NUM reserved the right to propose specific policies. A Labour government was *de facto* better than a Conservative one so 'It would be the greatest folly to permit feeling of dismay and any strained sense of loyalty to distort our judgement of the political issues involved at this election.'[86]

NUM negotiators strove to reconcile the 1966 Conference decision with support for Labour. They listed the factors which bargaining now had to take into account; government incomes policy, the industry's finances, and the TUC vetting committee and in the face of these restrictions it agreed to seek improvements in daywage rates and the continued implementation of NPLA.[87] The TUC had opposed prices and incomes legislation and preferred voluntary restraint but Part IV of the Prices and Incomes Act would lapse in August and would not be renewed. Ministers doubted the TUC could operate an effective voluntary policy and proposed to strengthen its delaying power by extending it from four to twelve months. The NUM's problem was that the 1966 Conference had explicitly rejected the early warning of claims but the government's supporters argued these powers were for a limited period only and if they were not accepted it was possible that the TUC would take over negotiations from individual unions. Those hostile to the policy argued that on the contrary, the General Council appeared increasingly resistant to government policy because of that policy's rejection of voluntarism, although both the General Council and the government were willing to accept productivity deals.

NPLA was by definition not a productivity deal.[88] The only way the pay circle could be squared was by drastically reducing the industry's costs to create a surplus to finance wage increases which meant 'Concentration into the most efficient pits and seams must obviously be accompanied by the maximum utilisation of machines and equipment available in the modernised collieries. This can involve the extension of multi-shift production and the possible extension or production into weekend shifts.' This would mean the end of the Five Day Week Agreement which remained a powerful symbol in the industry and there was, the NEC acknowledged, hostility in the pits to radical changes in work practices. The 1966 Conference's call for substantial increase for daywagemen had been blocked by the prices and incomes policy and NCB resistance, the January 1967 review of NPLA had also been delayed by pay policy and talks with the NCB on the 1967 wage claim only began on 13 April.[89] Sid Schofield (Yorkshire) put the mineworkers' frustrations into a historical perspective:

> if we had exploited the circumstances that prevailed in the fifties when the commodity that we produce had a seller's market, we could have demanded a much better deal for our members...We did not take advantage of the private enterprise philosophy because at that time we cherished the ideals of nationalisation...We did not realise, however, that notwithstanding the fullest co-operation that we, as miners, have afforded Governments and the [NCB] since the advent of nationalisation...when we could have joined the "rat race" and demanded a better deal for miners, forgetting all others.

The miners' reward? Uncertainty, disillusionment, frustration and 'There has been too long a period of restraint as far as a our lower-paid workers are concerned.'[90]

By the time of the 1968 NUM Conference discontent over pay, including NPLA, was leading to a serious situation in the NUM. Paynter reiterated the NUM's policy of improving the position of the low paid and narrowing the differential between daywagemen and those on NPLA but the prices and incomes policy was a major obstacle.[91] A further complication was that productivity, which had increased rapidly in previous years with the growth of mechanisation, was beginning to stagnate under NPLA. However, there was pressure for some of the benefits of the productivity increase to be passed on to the workforce. To forestall trouble the NEC requested the remission of all wage resolutions. Alf Hesler cautioned that promises made when NPLA was signed were not being delivered as its negotiators 'did not foresee the difficulty that was going to confront the Union by government legislation, by the freeze, and the [Fuel Policy] White Papers'.[92] Sid Schofield took this line of analysis further by arguing that NPLA, low pay, and pay policy were 'stretching the loyalties of our Members towards the Union' as well as the Labour government.[93] Lawrence Daly, who would soon take office as General Secretary, believed 'if we are going to ensure that we regain – not retain because we have lost it already – but regain the confidence of the lower-paid...then there will have to be...a substantial [wage] advance'.[94] A delegate from Nottingham, Bill Savage, was even more critical, attacking the NEC whose 'members come to this rostrum and generate a great deal of hot air about the daywagemen, and I sometimes feel we are going to be drowned in tears of sympathy. But I feel that in between Conferences the National Executive go to sleep'.[95]

The NCB's response to the NUM's claim was slow because 'they had been bound to observe the early warning system under the prices and incomes policy.' It offered a 15s (75p) per week increase for daywagemen. The NEC felt this was 'unsatisfactory' but it 'was the maximum which could be obtained by negotiations.' If the claim was pushed further the NEC feared many more jobs would be lost and 1968 was already the worst year for closures but this did not prevent some NEC members urging the to NUM take industrial action and there had been some unofficial strikes in some coalfields. To ease this pressure the NEC repeated its strategy of 1964 and called an individual ballot vote in the first week of November 1968 with an NEC recommendation to accept the NCB's offer.[96]

Comparison of the 1964 and 1968 ballots shows that the legacy of the post-1957 closures, the massive programme of closures carried out by the Labour government, the stresses and strains caused by the implementation of NPLA, and the frustrations caused by incomes policy was passivity. No NUM Area, Section or Group rejected the NCB's recommendation and the percentage of those who had voted against the NCB's offer in the 1964 ballot fell in all cases. In 1964 an average of 28.9 per cent of the coalfield Areas and 24.7 per cent of the Sections had voted against the NEC's recommendation to accept, but in 1968 these percentages were 13.5 per cent and 11.5 per cent respectively. In 1968 only some 15 per cent of mineworkers were willing to take an action which *might* have culminated in a ballot on industrial action. An alternative interpretation is that mineworkers perceived the proposed action as too weak whereas a stronger recommendation could have elicited more support. The available evidence does not support this interpretation and the Areas which had suffered most from pit closures – Durham, Scotland and South Wales – registered the largest falls in the percentage voting against the NCB's offer.

Pro-Nuclear = Anti-Coal?

In February 1955 the Conservative government announced a nuclear power programme of 12 stations with a capacity of 1.5-2 GW to be built over ten years. This was trebled in March 1957 and 5-6 GW of nuclear capacity was planned for 1965 but in June 1960 this was scaled down. In 1961 a technical, economic and political crisis over nuclear power began which was not resolved until 1965 and the Labour government's adoption of the AGR reactor. Originally, nuclear power was seen as a supplement to coal and the 1955 decision was based on 'the fear of a gap developing between Britain's rapidly growing energy requirements and the capacity of her coal industry.'[97]

Ministerial decision making relied on information provided by (non-expert) civil servants who drew on Atomic Energy Authority (AEA) and CEGB advice. The AEA was responsible not only for advising ministers on nuclear technology and reactor choice but also for reactor research and development.[98] Nuclear power was attractive because it was vulnerable to neither Arab oil producers nor the NUM and was therefore a guaranteed secure bridge for any energy gap. In addition, as the energy of the future a nuclear programme promised access to lucrative export contracts. Perhaps as important was the belief that nuclear power was a powerful symbol of a modern post-Suez Britain, a Britain which had lost its Empire but which, Wilson pledged, would be forged anew in the white heat of a technological revolution. Nuclear technology would show Britain remained a world player alongside the USA and USSR.[99] George Brown saw the pit closure programme as of wider political significance as it would demonstrate Labour's willingness to govern and take harsh decisions which directly affected its core supporters but which were necessary in the national interest.[100] Some still thought it 'extraordinary that a Labour government should be so anti-coal and willing to swallow any propaganda for atomic energy'. The NCB's Scientific member, Leslie Grainger, who had risen through the AEA to become assistant director at Harwell, believed 'the reason [for the nuclear programme] was to show that the Labour party is indeed in its element in the latter part of the twentieth century' while Robens ascribed the nuclear obsession not to the prospect of cheap energy 'but because it gave their people an immense lift to tell the world at power conferences that we produced more power from atomic stations than the whole of the rest of the world put together.'[101] How could coal, the quintessence of the first industrial revolution, and the NUM, the epitome of Old Labour, possibly compete? So, 'In short, the AEA told the Government that the AGR was what it was looking for and the government joyfully acceded to the charade.'[102] Even so forceful and skilled a Whitehall operator as Robens found it impossible to challenge the nuclear presumption which was reinforced by the closed nuclear network and the Ministry's culture of secrecy. Throughout 1967 and 1968 Robens tried repeatedly to secure internal data from the Ministry which would show at what level coal would be competitive with nuclear power. Faced by a direct request Richard Marsh saw no problem in providing the data but several weeks later Robens received a letter directing him to the columns of Hansard rather providing than the detailed statistics he wanted.[103]

The powerful Whitehall nuclear lobby meant 'there was no such thing as a Conservative or a Labour reactor. Certainly the expansion of nuclear power accorded with the main policy goal of both major parties'.[104] This was reflected in Fred Lee's May 1965 statement on the superiority of the AGR reactor. Lee's statement was based on the Dungeness B Appraisal of AGR and PWR tenders submitted to the CEGB and jointly assessed with the AEA which found the AGR had technical advantages over the PWR and would generate base-load electricity 10 per cent cheaper than Drax in West Yorkshire, the most modern coal-fired plant then under construction. In one of the most absurd ministerial pronouncements ever made Lee declared this to be 'the greatest breakthrough of all time.'[105] Hyperbole flowed from the minister's inability to challenge the pro-nuclear consensus or as Lee expressed it, 'I have consulted the AEA, the CEGB and the Ministry's Chief Scientist's Division. The advice from these totally different angles was unanimous. Obviously, they all could be wrong...As a non-technical Minister I can only say that in my judgement this seems highly unlikely.'[106] In a later debate Shinwell cautioned Marsh that in his experience it was dangerous to rely on civil service advice; 'I cannot help reflecting on how often I was provided with wrong estimates. I remember that on one occasion I gave the House an estimate of electricity consumption based on what the experts had told me. What did I know? ...I listened to the experts, only to be told a week afterwards they had given me the wrong estimates.'[107] The Ministry was in no position to dispute the AEA's analyses so its 'civil servants readily accepted this situation...incorporating [them] into plans, programmes and estimates, because that was what government policy required. They appeared to believe that all AEA judgements and assessments were right, all critics of the AEA wrong, and all AEA answers to criticisms right.'[108]

Robens complained vociferously about the effect on policy of frequent changes of ministerial personnel and philosophy:

> I have had [since 1961] four Ministers, three Parliamentary Secretaries; Ministers who have moved from wanting a business agenda to others who have wanted only a small, light chat, and Permanent Secretaries who have taken a different view about their own position in relation to the industry.[109]

This contrasted starkly with the stability of the nuclear network. Robens blamed much of the industry's difficulties on Wilson's appointment of Lee, who knew nothing about energy, as Minister of Power instead of Tom Fraser, the Opposition energy spokesman who was appointed Transport minister. Wilson's appointment of Lee:

> turned out to be a complete mistake. He never had a chance, for his background knowledge, experience and training had not equipped him to be the head of an economic ministry...he was square peg put in a round hole by a man [Wilson] who should have known better.[110]

Crossman saw Lee's speedy and uncritical adoption of the departmental line as the inevitable consequence of this lack of preparation. 'It would', he wrote, 'have been difficult to conceive nine months ago that Mr Lee would have been opposed to any help for the coalminers and blind to the fact that tapering subsidies [to ease the rundown] are politically essential.'[111]

Lee's replacement, Richard Marsh, was the epitome of Wilson's New Labour. Crossman described him as 'a tall, willowy young man with silk suits and mauve ties' who despite his trade union background 'mixes very naturally with the wealthy'; he was 'young and brash and devoid of any settled convictions' with a tendency to follow uncritically the departmental brief.[112] Marsh also complained 'that he is expected to make decisions on important matters when his knowledge is only what could be put on the back of a cigarette card. He then has to co-ordinate the policy of his ministry with other ministers, whose knowledge is even less.'[113] Crossman had speculated on the reasons for Lee's harshness towards the NUM; 'Is it because he is working class or is it because he is in the hands of the officials?' History, he felt, repeated itself with Mason. 'I noticed', Crossman wrote, 'that Roy Mason is already firmly against more coal-fired stations. It's astonishing how quickly these working-class boys get taken over by the civil servants.'[114] Robens dismissed Mason as 'just another minister who came into office with the right basic ideas and then allowed his judgement to be undermined by the civil service advisers.'[115]

In his memoirs Mason comments on the hostility of the head of the ministry's Coal Division to the industry and though he managed to move him sideways, he could not change the basic policy: 'I hated what was happening, fretting in my office as death sentences were passed on this and that pit. I was beginning to learn all about the limits on ministerial power.' The chief beneficiary was nuclear power. The best that Mason could achieve was to bargain power station licenses with the CEGB who in return for one oil and one nuclear station offered Mason Drax 2 which with Drax 1 became the largest coal-fired power station in Western Europe.[116] In 1971 Robens wrote that 'The idea that Britain should shed most of its coal industry, with all its attendant problems, was widespread in the Civil Service', citing James Callaghan's (then Chancellor of the Exchequer) opinion that every Treasury civil servant favoured the pit closure programme. This general anti-coal sentiment 'represented the thinking of the permanent officials and advisers of the various Ministers' which meant the NUM was consulted only when all the major decisions had been taken.[117] Neither the NCB nor the NUM were able to shift the policy or even insert any arguments into the decision-making process which conflicted with the departmental view.[118]

The power of the nuclear lobby was demonstrated by the government's decision to build an AGR nuclear station at Seaton Carew on the fringe of the Durham coalfield. This was one of the issues raised with Wilson at the 1968 Durham Miners' Gala. The miners received a sympathetic hearing and 'came away fully confident that the Prime Minister and his colleagues had finally recognised the force of their case.' At the same time Wilson gave the impression to his colleagues that he was willing to approve a coal-fired station at Seaton Carew to placate the NUM.[119] Paynter was furious:

> it is damn stupid on the part of any Government to conceive of developing nuclear power within the periphery of a coalfield. It is asking for trouble. It is inviting hostility and opposition from a section of the community in Britain that has been loyal, especially to this kind of Government'.[120]

In response to the furore Marsh called for a review of the CEGB's proposals. Before this was completed Marsh was replaced by Ray Gunter in a reshuffle on 6 April 1968 but, tired and emotional, Gunter resigned and was replaced by Roy Mason, the Postmaster General and an ex-miner and NUM sponsored MP for Barnsley on 1 July 1968. Invited for lunch with Wilson at Chequers, Mason claimed to have 'smelled a rat' and pointedly ask Wilson if there were any 'skeletons in the cupboard' which he denied. Mason remained suspicious as he was aware of both Wilson's duplicity and the problems at the Ministry of Power. Mason's instincts proved correct and the Seaton Carew decision, he claims, led him to contemplate resigning by mid-afternoon on his first day at the ministry. Mason protested to Wilson who, he claimed, did not appreciate the sensitivity of the decision and Mason tried to persuade him to take the decision back to Cabinet. Wilson refused on the grounds that the decision had been taken and even senior Labour politicians in the North East accepted the decision.[121]

The decision was announced in July when Parliament was in recess. Robens commented that 'The effect on the industry's morale was shattering' and 'The miners and the management began to despair of ever getting fair treatment from any Government.'[122] Crossman commented that 'We are so deeply convinced that, despite heavier capital costs, its running costs would be vastly less than a coal-fired station that we've insisted on building on top of a coalfield and creating unemployment among the miners.' He confessed to a gut feeling that the Seaton Carew decision was wrong but lacked the expertise to challenge it.[123] The Ministry of Power refused the coal industry's call for an independent inquiry into the AGR's costs with, according to Robens, Mason refusing on the grounds that the comparison would be so unfavourable to coal that the industry's remaining confidence would be shattered.[124] Although the Board had been kept informed about Seaton Carew, neither the NCB nor the government thought it proper to inform the NUM about the proposal even though they had met Mason on 12 July to discuss pit closures. The NUM was 'disgusted' but concluded fatalistically there was no point in trying to change the Cabinet's decision.[125] To Mason Seaton Carew 'felt like a betrayal', a judgement echoed throughout the NUM:

> I was a former miner. I remained a member of the NUM. I knew the anger and despair that would be felt in the pits when it was learned that coal was to be spurned in favour of a nuclear station. And what would certainly add insult to injury was the fact that the thing was to be built on top of a mountain of coal.[126]

Seaton Carew went ahead but in an effort to avoid a repetition Mason resolved to use his licensing powers to block any further nuclear power stations.

The NUM secured access up to and including the Prime Minister but its access to 'political administrators and their community relationships' was very restricted.[127] The NUM had no consistent or coherent direct contacts with the fuel policy network centred on the Ministry of Power and any residual influence was articulated via the NCB which did have these contacts but was itself in a subordinate position. Access was not influence:

> for eight years we have been running around in circles, knocking on doors, and the doors have always opened. Oh, yes, they always open, you are always welcomed in and you are sat down and there is a cup of tea provided. We have had meetings with Robens and his Board, we have had meetings with Marsh, we have had meetings with Wilson, not one but many meetings, and at the end of the road we are in the same position now as we were eight years ago...[128]

The policy style in the fuel policy network was closed and the creation of an Energy Advisory Council (January 1965) was symbolic and had no effect on energy policy.[129] By the summer of 1968 some ministers were beginning to doubt the nuclear power policy and pit closure programme.[130] By the winter of 1970/71 there were coal shortages in some areas of the country.

The NUM's Political Exclusion

In 1968 the coal industry had been nationalised for just over twenty years. During this time it had lurched from crisis to crisis, it had failed to meet many of the ideals of the founding generation but mineworkers' conditions had improved markedly for the better and the industry had been modernised and mechanised. The NCB was part of the post-war state but the NUM's relationship with both the NCB and the state was characterised by exclusion rather than inclusion.

To understand the NUM's exclusion we have to recognise that a political system is divided into a 'surface' and a 'deep' polity. The surface polity refers to the formal, visible political processes and institutions of elections, political parties, interest groups and parliament; the deep polity refers to the administrative state which is autonomous from the surface polity. These polities overlap as ministers and civil servants are present in both but it is in the deep polity where core decisions are taken. The most visible political institutions handle only secondary issues. Liberal democratic institutions and politics 'provide a useful and complex institutional façade, which absorbs political energies in non-threatening ways while masking the effective centralisation of power.'[131]

Exclusion: the Deep Polity

The NCB was part of the state and carried out state functions. The first function is the *maintenance of order* which requires social expenditure (on, for example, police) but in para-state organisations such as the NCB maintaining order took the form of the creation of an extensive, rules based systems of conflict management and consultation. In the coal industry these took the form of the consultation and conciliation structures organised in the CINC and JNNC. These operated in a highly pluralistic fashion but were heavily constrained by external political and economic influences which restricted NUM influence. The second function is *maintaining capital accumulation* by social investment on, for example, infrastructure. In the case of coal the nationalisation of a loss making but irreplaceable part of the country's infrastructure was a major instance of social investment. The third function is the manufacture of legitimation by *promoting social consumption* through, for example, public sector health and welfare services. In the coal industry this took the form of increased health, welfare and safety provision, improved pay and working conditions and the articulation of an ideology stressing co-operation between management and workforce with the object of transforming the industry's history and culture of conflict. The promotion of continued capital accumulation is the state's, and therefore the NCB's, most important task and this also the main concern of the deep polity. The preservation of order and the manufacture of consent and legitimacy are also of concern to the deep polity but these are the most visible concerns of the surface polity and the processes of liberal democracy.[132]

Despite the veneer of socialist rhetoric and expectation the Morrisonian public corporation was a technocratic and conservative organisation compatible with and indeed essential to, the capitalist political economy. Nevertheless the state industries were significant in four respects. First, they were part of the response to the rise of labour and class politics and coal nationalisation sought to manage this by incorporating a highly significant segment of organised labour into the state. Class compromise and accommodation were particularly important considerations in the case of the coal industry because of the political turbulence it had caused in the first half of the century. Second, public corporations such as the NCB were not socialist islands in a capitalist sea liable to inundation but were run-down and unprofitable but they were nonetheless essential to the success of private industry. As private enterprise was unable to sustain the coal industry, the state shouldered the burden although this was complicated by the ideological and emotional baggage associated with public ownership in the labour movement and particularly amongst the mineworkers. Public ownership *per se* posed few doctrinal problems for the Conservative Party or Conservative governments. Third, whilst state industries were part of the liberal democratic polity they were once removed from representative institutions and were really part of the administrative–bureaucratic state. The linkage between the two polities was the minister. Corporations like the NCB were technocratic-managerial bodies responsive to the pull of the deep state. Fourth, within these industries the ideology and ethos of traditional managerial and business elites predominated. Senior NCB management was socially and culturally closer to decision makers in the deep state and became more so as the state–coal industry relationship evolved and orthodox business ethics and accountancy practices determined the governance of the coal industry.

Liberal democratic institutions, in which the NUM was prominent, did effect policy but in the dual polity the state responds more quickly to business interests.[133] The NUM was a highly visible presence on the Labour Party's NEC and the TUC General Council, and in the 1950s enjoyed the status as one of the 'Big Six' unions in the labour movement. Even when the NUM began to shrink it remained a powerful symbol and important voice in the labour movement.[134] The mineworkers were well represented in the liberal democratic polity but not at all in the polity dominated by civil servants and professional managers. Both these groups were well aware of the NUM's potential power and, after 1957, of its perception of powerlessness and these perceptions were built into their calculations about coal policy. The NUM remained subordinate. The NCB's financial difficulties and the bureaucratic networks of the deep state gave it considerable influence over Board policy without undermining publicly the NCB's responsibility for the industry by overtly challenging the arm's length principle of political control which was central to the Morrisonian public corporation. In its anxiety to defend the coal industry from 'political interference' and to avoid any situation in which the NUM might find itself confronting the state, the NUM colluded in this charade. The minister–department–board relationship was a sectoral corporatism in which overall policy was determined centrally, transmitted via the NCB and implemented jointly by the NUM and Board, an arrangement which suited both Labour and Conservative governments. These weak linkages were reinforced by government's refusal to lay down a coal-centred fuel policy and its decision to support and promote consumer choice in fuel use. In this way the NCB was isolated from the liberal democratic polity and the NUM excluded from core decision making in the deep state on, for example, investment planning. Whilst the NCB's core decision making was insulated from liberal democratic control, the resulting policies were open to pressure in the surface polity but despite its substantial presence in this arena the NUM's parliamentary–political resources were insufficient. Exclusion from the centres of core decision making and the NUM's refusal to contemplate industrial action for political objects further neutralised the NUM.

So despite the NUM's importance in the post-war political economy it was excluded from effective political influence. Meetings with ministers occurred only when there was a crisis in the industry and in these discussions the NUM was invariably on the defensive and was called in to help resolve them. At all other times union–state relations were mediated through the NCB which confirms the political importance of the consultation system organised around the JNNC and the consultation machinery culminating in the CINC in helping exclude the NUM from significant influence. These two systems were deliberately kept separate and so helped fragment the NUM's potential influence: in the CINC system the NUM was expected to cooperate with the NCB in developing the industry whereas in the JNNC there remained two distinct and potentially opposed sides. The consultation and conciliation systems combined were designed to confine coal industry politics to the coal industry and helped prevent them spilling over into the wider political system.

The coal industry was a tripartite structure in that ministers, managers and workers were involved in the industry's processes but there was no tripartite forum where all three met together. The minister–board relationship although largely informal, was infinitely more institutionalised and regular than the minister–union relationship, and the strongest organisational relationship in the coal industry was between management and union. The coal industry was a *bifurcated tripartism* in which core decisions on investment, planning, prices and restructuring remained under managerial control within the administrative state with the NUM confined to symbolic consultation and the legitimisation and implementation of managerial decisions. The NUM enjoyed access not influence, status rather than power. Worker and union involvement in the affairs of 'their' industry was severely circumscribed and controlled by structures and procedures which, it has to be emphasised, the NUM advocated.

In his classic critique Panitch defines liberal corporatism as 'a political structure within advanced capitalism which integrates organised socio-economic producer groups through a system of representative and co-operative mutual interaction at the leadership level and of mobilisation and social control at the mass level'.[135] This describes the politics of the nationalised coal industry down to the late 1960s and demonstrates the profound weakness of 'corporatist' politics in post-war Britain. It testifies to the failure, indeed inability, of the Morrisonian public corporation to resolve management–worker conflict and which exacerbated union–worker conflict because of its emphasis on unions as managers of discontent. In the coal industry state, management and union relations were dominated by traditional collective bargaining issues and attitudes rather than characterised by the concertation of social partners. Whatever its intentions the NCB was incapable of meeting the expectations placed upon it by its employees and so like its private predecessor it remained a source of turbulence in the political economy.

Exclusion: the Surface Polity

At the 1967 Labour Party Conference the NUM had moved Composite Resolution 22 which called for an energy policy based on the maximum use of indigenous fuels, that greater aid be given to the coal industry to facilitate its transition, and more effort be devoted creating jobs in the coalfields. The NUM did not object to the creation of a 'virile, streamlined coal mining industry' but its motivation for this resolution was a 'deep rooted fear' that Labour was following a policy which disregarded previous party conference decisions.[136] The resolution was seconded by Sunderland North CLP whose delegate was less diplomatic warning that the government's policy was eroding 'the backbone of the Labour Party'. He continued, 'You have 1,000 people living in a village, so get them some employment in the village before you close the pits. It is no good taking a miner and getting him to sew knickers.' Party strategists talked constantly about the importance of the marginal seats whilst ignoring safe areas like the North East coalfields 'that put Harold Wilson and the rest where they are today.'[137]

Much of the criticism of Labour was related to Labour's failure, or reluctance, to challenge the deep state which led to Labour abandoning its beliefs. Arthur Palmer, the MP for Bristol Central, complained that 'two successive Labour Ministers of Power have endorsed the basically private enterprise system' of the Conservatives. He thought it 'extraordinary how policies which we said were against the public interest when they were put forward by Conservative Ministers become administrative common sense in the pure hands of Labour Ministers.' This flowed from the 'departmental assumption – that private enterprise is normal and public ownership is something special and difficult.' A delegate from the Rhondda West CLP condemned the government's uncritical acceptance of the perception 'that this is a declining industry and there is not very much we can do about it.'[138] Replying for Labour's NEC, Ian Mikardo reiterated the view that whatever the problems and difficulties a Labour government should be different and 'One of the reasons why we nationalised the coal industry in the first place was to get away from the situation which existed under the private employers when coal mining areas could be devastated without taking any of the social consequences into account.'[139] Ministers argued that they had taken these social consequences into account and that no opponent of the restructuring programme had shown how the unwanted coal could be used other than to advocate 'socialist planning'.

This fascinating debate encapsulated the difficulties of Labour in government, expressed the tensions inherent in Labour's modernisation ethos, and revealed the weaknesses of the party–union relationship. Whereas the NUM was excluded from the deep state it was a visible actor in the surface polity and had significant resources to deploy but it was only one of many interests. In the case of the party–union relationship the Wilson government relied on the doctrine of parliamentary sovereignty to insulate itself from external pressure and between 1966 and 1969 the government ignored 12 major Conference defeats. After the 1967 Conference the NEC tried to create a workable compromise, but a meeting between the Home Policy Committee and the Minister was a simply a presentation of each side's case and nothing was achieved. The government did not change its policy and 1968 was the worst year for closures. Ministers appeared not only insensitive but dismissive, a response which led to the NUM to openly oppose government policy at the 1968 Conference.

The Report on Fuel Policy presented to the 1968 Conference had been designed to maximise agreement but it backfired badly. At the 1968 Party Conference the NUM mobilised its resources (the chair of the Conference Arrangements Committee was a miner) and presence to ensure that a vacant Conference slot was allocated to the NUM to move the reference back of the section of the NEC's Report on fuel policy as part of the debate on energy policy. Many Conference delegates were influenced by the intrusion of a small group of protesting mineworkers during the Chair's Address as well as a general desire to exploit an opportunity to register a protest at much of what the Wilson government was doing. This produced a spectacular defeat which condemned the government's failure to act on the 1967 resolution, a defeat which was ignored by Ministers.[140] A meeting between Paynter, Ford and Bullough and representatives of Labour's NEC with Wilson and the relevant ministers at Downing Street on 7 May 1969 achieved nothing. There were no further meetings and no debates at the 1969 and 1970 Labour conferences on the coal situation.[141]

The inability of the NUM to influence policy through its sponsored MPs, structural links with the Labour Party, and by interest group action encouraged a growing scepticism amongst some mineworkers about the utility of political action via the parliamentary system. The NUM's dilemma can be seen in the 1970 General Election when it urged mineworkers to vote Labour to keep the Conservatives out, to preserve Labour's social legislation, because of Labour's aid to the industry, and because 'Experience has shown that we shall obtain a more reasonable solution from Labour than from a Tory administration.'[142] The NUM's disappointment with the Wilson government was counter-balanced by its history of loyalty to Labour and animosity to the Conservative Party:

> One would think it fair to assume that even the critics would accept that the Labour Government has gone a long way to try to minimise any hardship on our members...I honestly believe that if we had a Conservative Government...the contraction of our industry would have been much more savage, and the victims – our members – would not have received the same assistance to help cushion the effects of unemployment.[143]

In the 1970 election there was no evidence of a swing against Labour in mining constituencies.[144] An analysis of census data and electoral behaviour shows that the constituencies most likely to vote Labour in 1970 were the mining and heavy manufacturing of South Wales, the North-East, West and South Yorkshire, Derbyshire and Nottinghamshire and in the 50 most mining seats Labour voting held up. Labour paid no electoral price even in those coalfields which had suffered most from pit closures but the NUM *was* undergoing rapid political change. This change was reflected in the new legitimacy of industrial action in the unofficial strikes of autumn 1969 which were a reaction to the events explored above.[145] This change was to have profound consequences for British politics.

Notes

1. R.H.S. Crossman, *The Diaries of a Cabinet Minister. Volume 1, Minister of Housing 1964-1966*, Hamish Hamilton & Jonathan Cape 1975, p. 352, 18 October 1965. Hereafter *Diaries Volume 1*.
2. Memorandum by the Secretary of State for Economic Affairs, 29 June 1965. *CAB129/21*.
3. Situation in the Industry, *NUM(EC)*, 25 November 1965.
4. 5s H.C. Debs 721, 25 November 1965, col. 784.
5. 5s H.C. Debs 721, 25 November 1965, col. 829.
6. Appendix I. Economic Sub-Committee 8 December, *NUM(EC)*, and Meeting with the Minister, *NUM(EC)*, 9 December 1965. Crossman implies that by the end of 1965 ministers were fed up with the coal industry and pit closures. They recognised their political, social and economic importance but felt they were being diverted from other more pressing issues. *Diaries Volume 1*, p. 399, 2 December 1965.
7. Government White Papers on Fuel Policy and the Finances of the Coal Industry, *CINC (112)*, 13 December 1965. See the correspondence between Robens and Lee in the summer of 1965 in *POWE 52/72*.
8. R. Marsh, *Off The Rails: An Autobiography*, Weidenfeld and Nicolson 1978, pp. 111-12 for ministerial disquiet with Robens. Marsh was Minister of Power from October 1966 to April 1968.
9. Meeting with the Minister, *NUM(EC)*, 13 January 1966.
10. NUM, *Report of a Special Conference*, 18 February 1966, pp. 13-15.
11. *Report of a Special Conference*, p. 24, p. 27, p. 28, and p. 30.
12. Meeting with the Minister, *NUM(EC)*, 12 May 1966. In his memoirs Marsh writes, 'Did I interfere? Of course I did. I sought to apply back-stage pressures and did all the things ministers do.' *Off The Rails*, p.105.
13. HC 371-II. Select Committee on Nationalised Industries 1967-68, *Ministerial Control of the Nationalised Industries*, July 1968. *Minutes of Evidence*, q. 514. Hereafter *HC371-II*.
14. Marsh, *Off The Rails*, p.116.
15. *A Future for Miners*. Talking Points No.2, The Labour Party, 1966, p. 2 and p. 6.
16. HC 371-II. *Relations with the Treasury and Ministry of Power*. Memorandum submitted by the National Coal Board. No. 22, March 1967.
17. Appendix IV. Minutes of a Meeting between Economic Sub-Committee and the Minister of Power, 28 June, *NUM(EC)*, 1 July 1966.
18. NUM, *Report of the National Executive Committee*, May 1966, p. 9.
19. *NUM (ACR) 1966*, p. 91. The main issue at the Conference was not closures but incomes policy and this will be discussed later. Wilson was to have addressed the Conference but withdrew because of the economic crisis and the Seamen's strike.
20. *NUM (ACR) 1966*, p. 143.
21. Situation in the Industry. Letter from the Minister of Power, 8 December 1966, *NUM(EC)*, 11 January 1967.
22. Appendix I. Economic Sub-Committee, 8 February, *NUM(EC) Special Meeting*, 1 March 1967. A Policy for Coal, 14 February 1967, *POWE 52/11* emphasised the industry's continued decline to 150m ton (1970) and 100m ton (1980). A high level of state aid would be required to mitigate the social and economic consequences of decline.
23. Situation in the Industry, *NUM(EC)*, 11 May, and Meeting of the NEC with the Miners' Parliamentary Group at the House of Commons, 1 June 1967.

24 Robens, *Ten Year Stint*, pp. 214-17 and Marsh, *Off The Rails*, pp.108-109. The conference papers can be found in *POWE 52/119*.
25 Meeting with the Minister, *NUM(EC)*, 8 June, and Situation in the Industry, *NUM(EC)*, 4 July 1967.
26 *5s H.C. Debs 749*, 4 July 1967, col. 1536, col. 1538, and col. 1542.
27 *NUM (ACR) 1967*, p. 74.
28 *NUM (ACR) 1967*, p. 182.
29 *NUM (ACR) 1967*, p. 230.
30 *NUM (ACR) 1967*, p. 240.
31 *NUM (ACR) 1967*, pp. 245-46, p. 248, and p. 249. Joe Gormley was a member of the Labour Party's National Executive Committee.
32 Situation in the Industry, *NUM(EC)*, 20 July 1967. Marsh insisted no decision on Seaton Carew had been made.
33 R.H.S. Crossman, *The Diaries of a Cabinet Minister Volume 2. Lord President of the Council and Leader of the House of Commons 1966-68*, Hamish Hamilton and Jonathan Cape 1976, p. 431. Entry for 18 July 1967. Hereafter Crossman, *Diaries Vol. 2*. Miners' MPs wanted a full debate but as the House was about to recess there was insufficient time. The Order sought to increase the NCB's borrowing powers by £50 million under the Coal Industry Act (1965).
34 *5s H.C. Debs 759*, 18 July 1967, col. 1862.
35 Crossman, *Diaries Volume 2*, p. 432. 18 July 1967. The debate lasted from 10.25 p.m. on 18 July to 8.12 a.m. on 19 July (cols. 1859-2036).
36 *5s H.C. Debs 759*, 18 July 1967, col. 1982.
37 Crossman, *Diaries Volume 2*, p. 431. 22 July 1967 and pp. 451-52. 31 July 1967, and Note of a Meeting held in Room 33 at the County Hotel Durham on Saturday, 15 July 1967. *PREM 13/1610*. See also, Meeting between the PM, Minister of Power and Chairman of the NCB, 20 and 27 September 1967. *POWE 52/75*.
38 Robens, *Ten Year Stint*, p. 169, Marsh, *Off The Rails*, p.105. Notes for PM's Meeting with Lord Robens, 20 September 1967, and Record of a Meeting between the PM and the NUM at Scarborough, 29 September 1967. *PREM 13/1610*.
39 Marsh, *Off The Rails*, p.105.
40 Meeting with the Prime Minister, Secretary of State for Economic Affairs and Minister of Power, *NUM(EC)*, 29 September 1967. The original proposal was to postpone 26 closures but Marsh would go no further than 16. Wilson described these proposals as 'wonderful'. Crossman, *Diaries Volume 2*, p. 479. 15 September 1967. See also, R. Marsh and J. Callaghan, 1 August 1967. *PREM 13/1610*.
41 Situation in the Industry, *NUM (EC)*, 12 October and 7 November 1967. Also, Appendix II. Economic Sub-Committee, Minutes of Meeting with the Minister of Power, 6 November 1967.
42 *Fuel Policy*, Cmnd.3438, November 1967, para 89.
43 *Fuel Policy*, paras 102-103.
44 *Fuel Policy*, para 113.
45 *Fuel Policy*, Appendix I, para 19.
46 M.V. Posner, *Fuel Policy. A Study in Applied Economics*, Macmillan 1973, p. 303.
47 Posner, *Fuel Policy*, p. 304. Posner had been part of the econometric modelling team at the Ministry of Power.
48 *The Finances of the Coal Industry*, Cmnd. 2805, November 1965 and *Nationalised Industries. A Review of Economic and Financial Objectives*, Cmnd. 3437, November 1967, para 5.

49 Crossman, *Diaries Volume 2*, pp. 571-72, 14 November, and p. 574, 15 November 1967. Kelley was left-wing Labour MP for Don Valley and an NUM sponsored MP, Joe Gormley was a right-wing member of Labour's NEC. See also Neil McBride to Jack Diamond, 20 August 1967 detailing bitterness towards Labour in South Wales which was boosting Plaid Cymru's support. *PREM 13/1610*.
50 Crossman, *Diaries Volume 2*, p. 584, 21 November 1967, and CC (67) 67th Conclusions, 21 November 1967. *CAB 128/42*.
51 Crossman, *Diaries Volume 2*, p. 586, 22 November 1967.
52 Situation in the Industry, *NUM(EC)*, 14 December 1967 and Economic Sub-Committee, 4 January 1968. A critical resolution of coal policy at the 1967 Labour Party Conference was simply ignored by Wilson leading several members of the NEC to argue that the NUM had to rely on its own resources to resist the closures.
53 Appendix I. Economic Sub-Committee, 7 February and Appendix II. Economic Sub-Committee, Meeting with the Minister of Power, 7 February, in *NUM (EC) Special Meeting*, 27 February 1968.
54 Appendix I. Economic Sub-Committee, 4 March. Meeting with the Prime Minister, *NUM (EC)*, 14 March 1968. Marsh and Peter Shore also attended the meeting at Downing Street.
55 White Paper on Fuel Policy, *CINC (122)*, 21 November 1967.
56 Situation in the Industry: Special Conference, *NUM (EC)*, 14 March 1968.
57 NUM, *Report of a Special Conference*, 15 March 1968, p. 28.
58 *Report of a Special Conference*, p. 23.
59 *Report of a Special Conference*, p. 30 and p. 34.
60 *Report of a Special Conference*, p. 19.
61 *Report of a Special Conference*, p. 25.
62 *Report of a Special Conference*, p. 25.
63 *NUM (ACR) 1968*, pp. 214-16. Paynter was replaced by Lawrence Daly from Scotland whose election campaign was part of a wider strategy to bring about political change in the NUM. Central to Daly's campaign was the use of industrial action by the NUM. The 1968 General Secretary election will be examined in Chapter 1. Volume 2.
64 *CINC (126)*, 29 October, and Economic Sub-Committee, 3 December 1968. Robens, *Ten Year Stint*, p. 229. The NUM's National Officials had met Roy Mason on 6 November to discuss the Sunningdale meeting. R. Mason, *Paying the Price*, Robert Hale 1999, p. 104. Mason became the minister in July 1968.
65 Wages. Meeting with the NCB, *NUM (EC)*, 21 April 1964 and, Wages, *NUM (EC)*, 23 April 1964.
66 NUM, *Report of a Special Conference*, 29 April 1964, p. 25.
67 NUM, *Report of a Special Conference*, 29 May 1964, pp. 8-12.
68 *Report of a Special Conference*, p. 29.
69 *Report of a Special Conference*, p. 39 and Wages - Ballot Vote, *NUM (EC)*, 23 July 1964.
70 During this period considerable animosity developed between the coalfield Areas and COSA (the white-collar section) which was seen as a major obstacle to improved conditions. These were not, after all, 'proper miners' and Ford was a member of COSA.
71 The best account is still L. Panitch, *Social Democracy and Industrial Militancy. The Labour Party, the Trade Unions and Incomes Policy 1945-1974*, Cambridge University Press 1975.
72 NPLA is critical in explaining political change in the NUM and for this reason NPLA is analysed in Chapter 1. Volume 2.

[73] NUM, *Report of a Special Conference*, 12 March 1965, p. 2 and p. 18, and Wages, *NUM (EC)*, 25 June 1965.
[74] *NUM (EC)*, 1 April 1965.
[75] *JNNC (150)*, 15 April, and Productivity, Prices and Incomes, *NUM (EC) Special Meeting*, 29 April 1965.
[76] Meeting between the NEC and NCB, *NUM (EC)*, 25 June 1965. Emphasis added.
[77] Wages, *NUM (EC)*, 9 December 1965.
[78] Claim for Increase in Wages, *JNNC (153)*, 15 December 1965 and (154), 3 February 1966, and Wages, *NUM (EC)*, 10 February 1966.
[79] NUM, *Report of a Special Conference*, 18 February 1966, p. 6 and p. 8.
[80] *Report of a Special Conference*, p. 11.
[81] Claim for Increased Wages, *JNNC (155)*, 24 March, and Overtime Ban, *NUM (EC)*, 25 March 1966.
[82] *NUM (ACR) 1966*, p. 97.
[83] *NUM (ACR) 1966*, p. 96.
[84] *NUM (ACR) 1966*, pp. 204-205.
[85] Prices and Incomes Standstill, *NUM (EC)*, 2 September 1966.
[86] Appendix. General Election Manifesto, NUM, *Annual Report of the NEC*, May 1967.
[87] *JNNC (Union's Side)*, 9 February 1967.
[88] Incomes Policy, *NUM (EC) Special Meeting*, 1 March 1967. NPLA was a daywage (rate for the job) payment system.
[89] NUM, *Report of the National Executive Committee*, May 1967, p. 9 and p. 13.
[90] *NUM (ACR) 1967*, pp. 123-24.
[91] *NUM (ACR) 1968*, p. 106.
[92] *NUM (ACR) 1968*, p. 117,
[93] *NUM (ACR) 1968*, p. 123.
[94] *NUM (ACR) 1968*, p. 126.
[95] *NUM (ACR) 1968*, p. 132.
[96] Wages, *NUM (EC)*, 10 October and 14 November 1968.
[97] R. Williams, *The Nuclear Decisions. British Policies, 1953-1978*, Croom Helm 1980, pp. 60-61.
[98] D. Burn, *Nuclear Power and the Energy Crisis. Politics and the Atomic Industry*, Macmillan 1978, pp. 168-75.
[99] For Labour's 1959 general election campaign Benn proposed a party political broadcast *Land and the People* whose 'opening film sequence should be an atomic power station, seen across fields of waving corn. And our music should be "Jerusalem", sung by a Welsh choir'. M. Cockerell, *Live from Number 10. The Inside Story of Prime Ministers and Television*, new edition, Faber and Faber 1989, p. 71.
[100] G. Brown to H. Wilson, 18 November 1965. *POWE 52/74*.
[101] C. King, *The Cecil King Diaries 1965-1970*, Jonathan Cape 1972, p.269 entry for 23 July 1969, and p. 312 entry for 4 February 1970.
[102] A. Massey, *Technocrats and Nuclear Politics. The Influence of Professional Experts in Policy-Making*, Avebury 1985, p. 119.
[103] *HC371-II*, para 508.
[104] Massey, *Technocrats and Nuclear Politics*, p. 119. See also E.P. (64) 51, 9 April 1964. Nuclear Power Programme. Memorandum by the Chancellor of the Exchequer. *CAB 129/117*. This set the broad parameters of the second nuclear power programme but did not decide in favour a particular reactor type.
[105] *5s H.C. Debs 713*, 25 May 1965, col. 238.

[106] 5s H.C. Debs 759, 18 July 1967, col. 1865.
[107] 5s H.C. Debs 759, 18 July 1967, col. 1886. For a similar view of the civil service's use of statistics see, R.H.S. Crossman, *Diaries of a Cabinet Minister. Volume 3, Secretary of Sate for Social Services 1968-1970*, Hamish Hamilton and Jonathan Cape, 1977, pp. 268-69 entry for 8 March 1967.
[108] Burn, *Nuclear Power and the Energy Crisis*, p. 176.
[109] HC. 371-II. Minutes of Evidence, q. 510. Between 1961 and 1971 Robens had a total of ten ministers.
[110] Robens, *Ten Year Stint*, p. 158.
[111] Crossman, *Diaries Volume 1*, p. 258 entry for 28 June 1965. See also p. 352, entry for 18 October 1965.
[112] Crossman, *Diaries Volume 3*, p. 81 entry for 27 May 1968.
[113] King, *Diaries 1965-1970*, p. 135 entry for 29 July 1967.
[114] Crossman, *Diaries Volume 1*, p. 352, entry for 18 October 1965, and *Diaries Volume 3*, p. 162, entry for 30 July 1968. The office of Minister of Power was abolished on 6 October 1969 and its responsibilities transferred to the Ministry of Technology.
[115] Robens, *Ten Year Stint*, p.172.
[116] Mason, *Paying the Price*, p. 104 and p. 106.
[117] Robens, *Ten Year Stint*, pp. 171-72, and Situation in the Industry, NUM (EC), 11 May 1967.
[118] HC371-II, para 471. In six years Robens met Treasury civil servants only once without being accompanied by minders from the Ministry of Power.
[119] Robens, *Ten Year Stint*, p. 198, and Crossman, *Diaries Volume 2*, p. 676-77, 12 February 1968.
[120] NUM, *Report of a Special Conference*, 15 March 1968, p. 15.
[121] Mason, *Paying the Price*, pp. 99-100.
[122] CINC (122), 21 November 1967, and Robens, *Ten Year Stint*, p. 199.
[123] Crossman, *Diaries Volume 3*, p. 162 entry for 30 July 1967. The coal-nuclear cost debate was complex, absorbed much time and produced much Whitehall in-fighting. See, for example, New Power Stations (particularly Seaton Carew) and Relative Costs of Electricity Generation by Nuclear Fuel and Coal. Powe 52/381.
[124] Robens, *Ten Year Stint*, p.173.
[125] Seaton Carew Power Station Decision, CINC (125), 27 August 1968. Robens blames Mason for being 'responsible for the biggest blow ever inflicted on the miners' morale'. *Ten Year Stint*, p. 57.
[126] Mason, *Paying the Price*, p. 100.
[127] H. Heclo and A. Wildavsky, *The Private Government of Public Money. Community and Policy inside British Politics*, Macmillan 1974, p. 2. Political administrators are the ministers and civil servants who preside over departments and Cabinet. Any individual or group outside this assemblage, is by definition, marginal.
[128] NUM, *Report of a Special Conference Report*, 15 March 1968, p. 24.
[129] 5s H.C. Debs 705, 22 January 1964, cols. 136-37. Robens and Ford were EAC members.
[130] Crossman, *Diaries Volume 3*, p. 124 entry for 8 July 1968.
[131] P. Dunleavy and B. O'Leary, *Theories of the State. The Politics of Liberal Democracy*, Macmillan 1986, p. 178.
[132] This paragraph draws on J. O'Connor, *The Fiscal Crisis of the State*, St Martin's Press 1973, A. Wolfe, *The Limits of Legitimacy*, The Free Press 1977, and A. Cawson and P.

Saunders, 'Corporatism, Competitive Politics and Class Struggle' in R. King ed., *Capital and Politics*, Routledge and Kegan Paul 1983.

[133] C. Lindblom, *Politics and Markets. The World's Political-Economic Systems*, Basic Books 1977, pp. 172-78.

[134] An indication of the NUM's presence in the formal liberal democratic political system can been seen in the size of the Miners' Parliamentary Group:

	1945	1950	1951	1955	1959	1964	1966	1970
NUM	34	37	36	34	31	28	26	20
PLP	393	315	295	277	258	317	363	287
Per cent	8.6	11.7	12.2	12.2	12.0	8.8	7.1	6.9

[135] L. Panitch, The Development of Corporatism in Liberal Democracies, in G. Lehmbruch and P.C. Schmitter eds., *Trends Towards Corporatist Intermediation*, Sage Publications 1979, p. 123.

[136] *Labour Party Conference Report 1967*, p. 202.

[137] *Labour Party Conference Report 1967*, p. 203-204.

[138] *Labour Party Conference Report 1967*, p. 205 and p. 206.

[139] *Labour Party Conference Report 1967*, p. 209.

[140] *Labour Party Conference Report 1968*, pp. 278-81 and *National Executive Committee Annual Report 1968*, pp. 98-100.

[141] L. Minkin, *The Labour Party Conference*, Manchester University Press 1980, pp. 305-306.

[142] A General Election Message from the National Union of Mineworkers, *NEC Annual Report for 1970*, p. 21.

[143] Yorkshire Area NUM Annual Reports, *The Report of the General Secretary for 1969* (March 1970), p. 91.

[144] D. Butler and M. Pinto-Duschinsky, *The British General Election of 1970*, Macmillan 1970, p. 394.

[145] These unofficial strikes will be examined in the first chapter of Volume 2.

Bibliography

The main source of primary information on the NUM for this book are the records of the National Union of Mineworkers and in particular the minutes of the National Executive Committee, the verbatim reports of the Annual Conference and periodic Special Conferences, the Annual Reports of the NEC, and various editions of the NUM's Rulebook. These also include the minutes and reports of the Coal Industry National Consultative Committee (CINC) and the Joint National Negotiating Committee (JNNC). Sources used at the Public Record Office include Cabinet minutes, papers, and memoranda (CAB128 and CAB129), personal files of prime ministers (PREM), and various files from the Ministry of Power (POWE). Some collections of private papers were used, for example, the Headlam Diaries at the Durham Record Office and the Woolton Papers at the Bodleian Library, Oxford. Also used from the Bodleian were files from the Conservative Party Archive (CPA). Details of these sources are given in the notes at the end of each chapter.

Place of publication is London unless stated otherwise.

Addison, P. (1993), *Churchill on the Home Front 1900-1955*, Pimlico.
Allen, V. (1981), *The Militancy of British Miners*, The Moor Press, Shipley.
Ashworth, W. (1986), *The History of the British Coal Industry. Volume 5 1946-1982: The Nationalised Industry*, Oxford, Clarendon Press.
Baldwin, G.B. (1955), *Beyond Nationalization. The Labor Problems of British Coal*, Harvard University Press, Cambridge, Mass.
Bamburg, J.H. (1994), *The History of the British Petroleum Company. Vol. 2 The Anglo-Iranian Years*, Cambridge University Press, Cambridge.
Barnett, C. (1986), *The Audit of War. The Illusion and Reality of Britain as a Great Nation*, Macmillan.
Barnett, C. (1995), *The Lost Victory. British Dreams, British Realities 1945-1950*, Macmillan.
Benney, M. (1978), *Charity Main. A Coalfield Chronicle*, (first published 1946) E.P. Publishing, Wakefield.
Berle, A.A. and Means, G.C. (1947), *The Modern Corporation and Private Property*, New York, Macmillan.
Brivati, B. (1997), *Hugh Gaitskell*, Richard Cohen.
Bullock, J. (1972), *Them and Us*, Souvenir Press.
Burn, D. (1978), *Nuclear Power and the Energy Crisis. Politics and the Atomic Industry*, Macmillan.
Butler, D., and Pinto-Duschinsky, M. (1970), *The British General Election of 1970*, Macmillan.
Cairncross, A. (ed.) (1989), *The Robert Hall Diaries, 1947-1953*, Unwin Hyman.

Campbell, A., Fishman, N. and Howell, D., (eds) (1996), *Miners, Unions and Politics 1910-1946*, Scolar Press, Aldershot.
Chester, D.N. (1975), *The Nationalization of British Industry*, HMSO.
Church, R. and Outram, Q. (1998), *Strikes and Solidarity. Coalfield Conflict in Britain 1889-1966*, Cambridge University Press, Cambridge.
Coates, D. (1975), *The Labour Party and the Struggle For Socialism*, Cambridge, Cambridge University Press.
Cockerell, M. (1989), *Live from Number 10. The Inside Story of Prime Ministers and Television*, new ed., Faber and Faber.
Cockett, R. (ed.) (1990), *My Dear Max. The Letters of Brendan Bracken to Lord Beaverbrook, 1925-1958*, The Historian's Press.
Cole, M. (1949), *Miners and the Board*, Fabian Research Group. Research Series 134, Fabian Society/Victor Gollancz.
Coombes, B.L. (1948), 'One Year of Nationalisation', *Fortnightly Review*, January.
Court, W.H.B. (1951), *Coal*, HMSO/Longman Green & Co.
Crossman, R.H.S. (1975), *The Diaries of a Cabinet Minister. Volume 1. Minister of Housing 1964-66*, Hamish Hamilton and Jonathan Cape.
Crossman, R.H.S. (1976), *The Diaries of a Cabinet Minister. Volume 2. Lord President of the Council and Leaders of the House of Commons 1966-68*, Hamish Hamilton/Jonathan Cape.
Crossman, R.H.S. (1977), *The Diaries of a Cabinet Minister. Volume 3. Secretary of State for Social Services 1968-70*, Hamish Hamilton/Jonathan Cape.
Davis Smith, J. (1990), *The Attlee and Churchill Administrations and Industrial Unrest 1945-1955. A Study in Consensus*, Frances Pinter.
Dennis, N., Henriques, F. and Slaughter, C. (1969), *Coal Is Our Life. An Analysis of a Yorkshire Mining Community*, 2nd ed. (first published 1956) Tavistock.
Department for Economic Affairs (1965), *The National Plan*, Cmnd. 2764, HMSO.
Dunleavy, P. and O'Leary, B. (1986), *Theories of the State. The Politics of Liberal Democracy*, Macmillan.
Eldon Barry, E. (1965), *Nationalisation in British Politics. The Historical Background*, Jonathan Cape.
Evans, B.J. and Taylor, A.J. (1996), *From Salisbury to Major. Continuity and Change in Conservative Politics*, Manchester University Press, Manchester.
Foot, R. (1945), *A Plan For Coal. A Report to the Colliery Owners*, Mining Association of Great Britain.
Francis, H. and Smith, D. (1980), *The Fed. A History of the South Wales Miners in the Twentieth Century*, Lawrence & Wishart.
Fryth, J (ed.) (1993), *Labour's High Noon. The Government and the Economy*, Lawrence and Wishart.
Gibbon, P. (1988), 'Analysing the British miners' strike of 1984-5', *Economy and Society*, vol. 17, pp.139-94.
Goodhart, P. (1973), *The 1922. The Story of the 1922 Committee*, Macmillan.
Gormley, J. (1982), *Battered Cherub*, Hamish Hamilton.
Griffin, C. (1989), *The Leicestershire Miners. Volume III, 1945-1988*, NUM Leicester Area, Coalville.

H.M. Treasury (1961), *The Financial and Economic Obligations of the Nationalised Industries*, Cmnd. 1337, HMSO.
H.M. Treasury (1967), *Nationalised Industries. A Review of Economic and Financial Objectives*, Cmnd.3437, HMSO.
Handy, L.J. (1981), *Wages Policy in the British Coalmining Industry. A Study of National Wage Bargaining*, Cambridge, Cambridge University Press.
Harris, K. (1982), *Attlee*, Weidenfeld & Nicolson.
Harris, N. (1972), *Competition and the Corporate Society. British Conservatism, The State and Industry 1945-1964*, Methuen.
Heclo, H. and Wildavsky, A. (1974), *The Private Government of Public Money. Community and Policy inside British Politics*, Macmillan.
Heinemann, M. (1944), *Britain's Coal. A Study of the Mining Crisis*, Victor Gollancz.
Hennessy, P (1993), *Never Again. Britain, 1945-1951*, New York, Pantheon Books.
Horner, A. (1960), *Incorrigible Rebel*, MacGibbon & Kee.
House of Commons (1958), *Report of the Select Committee on Nationalised Industries*, HC.187, HMSO.
House of Commons (1966), *Second Report of the Select Committee on Nationalised Industries: Gas, Electricity and Coal Industries*, HC.77, HMSO.
House of Commons (1967-8), *First Report of the Select Committee on Nationalised Industries*, HC.371, HMSO.
House of Commons Debates (various), *Hansard*, HMSO.
Howell, D. (1989), *The Politics of the NUM. A Lancashire View*, Manchester University Press, Manchester.
Kelly, M. and Forsyth, D.J. (eds.) (1969), *Studies in the Coal Industry*, Oxford, Pergamon Press.
King, C. (1972), *The Cecil King Diaries 1965-1970*, Jonathan Cape.
King, R. (ed.) (1983), *Capital and Politics*, Routledge and Kegan Paul.
Kirby, M.W. (1977), *The British Coalmining Industry 1870-1946. A Political and Economic History*, Macmillan.
Labour Party (various years), *Report of the Annual Conference*.
Labour Party/Trades Union Congress (1936), *Coal: The Labour Plan*, Labour Party and TUC.
Lambeth, Lord Morrison of (1960), *Herbert Morrison. An Autobiography*, George Allen and Unwin.
Lee, W.A. (1954), *30 Years in Coal. A Review of the Coal Mining Industry under Private Enterprise*, Mining Association of Great Britain.
Lehmbruch, G. and Schmitter, P. (eds.) (1979), *Trends Towards Corporatist Intermediation*, Sage Publications.
Lindblom, C. (1977), *Politics and Markets. The World's Political-Economic Systems*, Basic Books, New York.
Lorwin, V.R. (1954), *The French Labor Movement*, Harvard University Press.
Macmillan, H. (1971), *Riding the Storm, 1956-1959*, Macmillan.
Macmillan, H. (1973), *At the End of the Day 1961-1963*, Macmillan.

Marsh, R. (1978), *Off the Rails. An Autobiography,* Weidenfeld and Nicolson.
Mason, R. (1999), *Paying the Price*, Robert Hale.
Massey, A. (1985), *Technocrats and Nuclear Politics. The Influence of Professional Experts in Policy-Making*, Avebury, Aldershot.
McCallum, R.B. and Readman, A. (1964), *The British General Election of 1945*, Frank Cass.
McCormick, B.J. (1979), *Industrial Relations in the Coal Industry*, Macmillan.
Middlemas, K. (1986), *Power, Competition and the State. Volume 1, Britain in Search of Balance, 1940-1961*, Macmillan.
Ministry of Fuel and Power (1945), *Coal Mining. Report of the Technical Advisory Committee.* Cmd.6610, HMSO.
Ministry of Fuel and Power (1952), *Report of the Committee on National Policy for the Use of Fuel and Power Resources*, Cmnd.8647, HMSO.
Ministry of Power (1965), *Fuel Policy*, Cmnd.2798, HMSO.
Ministry of Power (1965), *The Finances of the Coal Industry*, Cmnd.2805, HMSO.
Ministry of Power (1967), *Fuel Policy*, Cmnd.3438, HMSO.
Minkin, L. (1980), *The Labour Party Conference*, Manchester University Press, Manchester.
Moffat, A. (1965), *My Life with the Miners*, Lawrence & Wishart.
Montague-Browne, A. (1995), *Long Sunset*, Cassell.
Morrison, Herbert (1933), *Socialisation and Transport*, Constable.
Nabarro, G. (1969), *NAB 1. Portrait of a Politician*, Robert Maxell.
National Coal Board (1950), *Plan For Coal.*
National Coal Board (1955), *Report of the Advisory Committee on Organisation,* (The Fleck Report).
National Coal Board (1956), *Investing in Coal.*
National Coal Board (1957), *The First Ten Years.*
National Coal Board (1959), *Revised Plan For Coal.*
National Coal Board (various issues), *Annual Reports and Accounts.*
National Union of Mineworkers (1945), *The Miners' Case. An Answer to the Foot Plan*, NUM, March.
O'Connor, J. (1973), *The Fiscal Crisis of the State*, St. Martin's Press, New York.
Page Arnot, R. (1961), *The Miners in Crisis and War. A History of the Miners' Federation of Great Britain (from 1930 onwards)*, George Allen and Unwin.
Page Arnot, R. (1979), *The Miners. One Union, One Industry. A History of the National Union of Mineworkers 1939-1946*, George Allen and Unwin.
Panitch, L. (1975), *Social Democracy and Industrial Militancy. The Labour Party, the Trade Unions and Incomes Policy 1945-1974*, Cambridge University Press, Cambridge.
Paynter, W. (1972), *My Generation*, George Allen and Unwin.
Pearce, C. (ed.) (1991), *Patrick Gordon Walker. Political Diaries 1932-1971*, The Historian's Press.
Pimlott, B. (1992), *Harold Wilson*, HarperCollins.
Pimlott, B. (ed.) (1986), *The Political Diary of Hugh Dalton 1918-1940, 1945-1960*, Jonathan Cape.
Posner, M.V. (1973), *Fuel Policy. A Study in Applied Economics*, Macmillan.

Ramsden, J. (1995), *A History of the Conservative Party. The Age of Churchill and Eden, 1940-1957*, Longman.
Ramsden, J. (1996), *A History of the Conservative Party. The Winds of Change: Macmillan to Heath, 1957-1975*, Longman.
Robens, Lord (1971), *Ten Year Stint*, Cassell.
Roberts, A. (1994), *Eminent Churchillians*, Weidenfeld & Nicolson.
Robertson, A.J. (1987), *The Bleak Mid-Winter. Britain and the Fuel Crisis of 1947*, Manchester, Manchester University Press.
Scott, W., Mumford, E., McGivering, I. and Kirby, J. (1963), *Coal and Conflict*, Liverpool University Press, Liverpool.
Seldon, A. (1981), *Churchill's Indian Summer. The Conservative Government 1951-55*, Hodder and Stoughton.
Seldon, A. and Ball, S. (eds.) (1994), *The Conservative Century*, Clarendon Press, Oxford.
Select Committee on Nationalised Industries (1967-68), *Ministerial Control of the Nationalised Industries*, HC.371, HMSO.
Shinwell, E. (1955), *Conflict Without Malice*, Odhams.
Slowe, P. (1993), *Manny Shinwell: An Authorised Biography,* Pluto Press.
Taylor, A.J. (1983), 'The Miners and Nationalisation', in *International Review of Social History*, vol. 28, no.2, pp.176-99.
Taylor, A.J. (1984), *The Politics of the Yorkshire Miners*, Croom Helm, Beckenham.
Taylor, A.J. (1996), 'Maximum Benefit for Minimum Sacrifice': The Miners' Wage Campaign, 1935-36', *Historical Studies in Industrial Relations*, vol. 1 no.2, pp.65-92.
Taylor, R. (1993), *The Trade Union Question in British Politics. Government and Unions since 1945*, Blackwells/Institute for Contemporary British History, Oxford.
Trades Union Congress (1944), *Interim Report on Post-War Reconstruction.*
Trades Union Congress (1948), *Communist Activities Examined.*
Trades Union Congress (1948), *Defend Democracy.*
Trades Union Congress (1948), *Warning to Trade Unionists.*
Trades Union Congress (various years), *Annual Report of Congress.*
Weiler, P. (1988), *British Labor and the Cold War*, Stanford, CA, Stanford University Press.
Williams, C. (1996), *Democratic Rhondda. Politics and Society 1885-1951*, University of Wales Press, Cardiff.
Williams, C. (1998), *Capitalism, Community and Conflict. The South Wales Coalfield 1898-1947*, University of Wales Press, Cardiff.
Williams, F. (1961), *A Prime Minister Remembers*, Heinemann.
Williams, P. (ed.) (1983), *The Diary of Hugh Gaitskell*, Jonathan Cape.
Williams, R. (1980), *The Nuclear Decisions. British Policies, 1953-1978*, , Croom Helm, Beckenham.
Wolfe, A. (1977), *The Limits of Legitimacy*, The Free Press, New York.
Zweig, F. (1948), *The Men In the Pits*, Gollancz.

Index

Aberfan 217
Accelerated Closure Programme 202-7, 213-15
AGR 218, 221, 223, 239, 240, 242-4
Allen, Bill 4, 140
Arthur, Will 42
Ashton study 91-2, 103
Ashworth, W.A. 114
Assheton, Ralph 26
Atomic Energy Authority 239, 240-41
Attlee, Clement 22, 24, 28, 45, 46, 47, 50, 51, 68, 94, 98, 100, 137, 198, 231

Ballot vote 6, 238-9
Barmoor Decision 71
BBC 134
Besford, Jack 59
betrayal thesis 197-207
Bevanism 137, 198
Bevin, Ernest 47, 51-2, 68, 69, 89
bifurcated tripartism 35, 247
Bowman, James 3,4 5, 6, 9, 45, 49, 63, 68, 71, 97, 100, 101, 156, 159, 164-6, 173, 192
Bracken, Brendan 153, 172
Braithwaite, Sir Arthur 130, 169
British Electricity Authority (*see* Central Electricity Generating Board)
Brook, Sir Norman 112-13, 119, 148-9, 172, 168, 171
Brown, George 201-2, 211-13, 232-3, 239, 254
Buchan-Hepburn, Patrick 130
Bullock, Jim 94, 97, 117, 205-6
Bullough, Sam 177, 250
Burke, Tommy 227-8
Burrows, S.R. 100
Butler, R.A. 111-13, 122, 128, 144-6, 148-9, 154

Callaghan, James 221-2, 242

Carr, Bill 218
Central Electricity Generating Board (CEGB) 148-9, 192, 221, 224, 229, 239, 240, 242-3
Challenge to Britain 198
Charter of Demands 43, 82-3, 88
Cherwell, Lord 126, 132, 135
Churchill, Winston S. 17, 18, 26, 121-5, 127-8, 131-2, 135-6, 142-4, 148-9, 151, 154
closures 43-4, 183, 184-94, 196, 198-9, 201, 203-7, 213-29, 231, 233, 235, 238, 239, 243, 249, 250,
coal crises 11-8 (1942), 41-53 (1947), 128-37 (1951-2), 144-51 (1954-5), 184-97-206 (1957-8), 202-7, 213-9 (1964-5), 219-29 (1967)
Coal Industry National Consultative Committee (CINC) 52, 85, 107, 187, 193, 205, 207, 226, 247
Coal Industry Nationalisation Act (CINA) 35-41, 54, 89, 108-9, 154, 193
coal industry reorganisation (*see* coal crises, closures) 11-20, 26-8, 151-9
Coal Is Our Life 92, 103
Coal Mines Act (1930) 16, 23
Coal Mines Regulation Act (1908) 49
Coal Mines Reorganisation Commission 16
coal policy (*see* fuel policy, coal crises, closures)
Coal. The Labour Plan 25
Cohen Council (*see* CPPI)
Cold War 61-73
collective bargaining (*see* wages, conciliation machinery)
Colville, Jock 164
communists (*see* Cold War*)* 3, 10, 55, 2-4, 67-9, 70, 73, 121, 127, 143, 162
conciliation machinery 28, 34, 38, 40, 44, 60-1, 82-93, 102, 104, 108, 115, 132, 139, 145, 153, 158,

161, 177, 181-2, 189, 214, 245, 247
Confederation General du Travail (CGT) 63-4, 65, 73
conference (NUM) 7-9
Conservative Governments
   and coal policy 17, 21, 25-8, 122-31, 184-97, 144-51
   and coal prices 108-15, 128-9, 134-6, 139, 141-2, 145-6, 148-51, 162, 166-7
   and Fleck Report 150, 151-9
   and industrial relations 127-9, 131-7, 175-84
   and miners' wages 131-7, 180-4
   and nationalisation 25-8, 122-31, 151-9, 184-97
   and the NCB 122-3
Conservative Party (*see* Conservative Governments)
Conservative Research Department 127
consolidationism 137-8
consultative machinery (*see* CINC) 3, 17, 37, 38, 40, 82-93, 94, 105, 158
Conway, Jim 59
Coombes, B.L. 101
corporate governance 37
CPPI 182-3
Crawford, Sir James 158
Cripps Plan 46-7
Cripps, Sir Stafford 46-7, 66
Crossman, R.H.S. 202, 213, 221-2, 225, 241-2, 243
cultural revolution 34-5, 42, 86, 93-108, 154, 245
Cumberland 5, 101, 102, 141, 206

Dalton, Hugh 18, 35, 45, 66, 197
Daly, Lawrence 238
daywagemen 111, 139, 140, 143, 161 180, 182, 232, 236-7, 238
deep polity 245-8
deep state 246, 247, 249
Degnan, Tommy 106
Derbyshire 90, 91, 133, 137, 159, 178, 214, 229, 230, 231, 250
disputes (*see* strikes)
Doncaster 83, 231
Dunn, Jack 216, 220
Durham 3, 6, 38, 49, 85, 95, 130, 132, 162, 206, 207, 216, 221, 222, 230, 231, 239, 242

East Midlands 146, 216
Eccles, David 167
Economist, The 20
Eden, Anthony 114, 123, 130, 151, 162, 164, 165, 166
Edwards, Ebby 25, 84, 86, 89, 100, 156
Ellis, Bill 70
Employment Policy (1944) 15
Extended Hours Agreement (*see* hours) 50, 161
Ezra, Derek 216, 229

Fabian Society 95, 99, 103
Fife Coal Company 12
Five Day Week Agreement (FDWA) 43, 44, 45, 47, 49-0, 53, 83, 88, 101, 102
Fleck Report 150, 151-9, 176, 191
Foot Plan 11, 17-21, 22, 26
Ford, Sidney 66, 204, 206, 218, 220, 222, 229, 230, 234, 235, 250
Francis, Dai 162
Fuel and Power. An Immediate Policy 201-2
Fuel Policy (1965) 214
Fuel Policy (1967) 223-4
fuel policy 11-12, 17, 19, 26-8, 34, 113, 128, 148, 184-5, 188-9, 190, 192-3, 198-206, 214-5, 218, 219-29, 232, 238, 244, 246, 249-50

Gaitskell, Hugh 35, 39, 40-41, 46, 49, 50, 56, 60, 63, 67, 73, 91, 94, 99, 100, 105, 124, 197
Gallacher, Willie 63
GEN 445 145-7
General Secretary (NUM) 10-11, 58, 64, 65, 66, 71, 84, 182, 238
General Strike, The 2, 13-14, 20, 26, 36, 47, 64, 94, 121, 127, 132, 134, 138-9, 144, 160, 162, 180, 189, 191, 226-8, 230, 234,
Gentlemen's Agreement 109-10
Gormley, Joe 98, 220-21, 225
Gray, Charles 123-4
Griffiths, Jim 51
Grimethorpe 83-5, 122
Group No. 2 231

Gunter, Ray 221, 243

Hall, Joe 21, 87
Hall, Robert 52
Hammond, Jim 42, 57, 70, 105, 137, 140, 167, 187
Haworth 4
Heathfield, Peter 205
Hesler, Alf 216, 227, 238
Horden 38
Horner, Arthur 4, 21, 38, 42, 46, 47, 48, 52, 55, 57, 58, 60, 61-73, 83, 88, 90, 91, 96, 100, 103, 104, 106-8, 110, 121, 133, 136, 138-9, 141, 142, 143, 157, 159, 160, 164, 177, 178, 179, 180, 182, 183, 188, 190
Houldsworth, Sir Hubert 52, 122, 133, 144, 145, 153, 154, 155, 156, 157, 161, 165
hours of work 41, 47-8, 50, 85, 133, 139, 162-5, 167, 195
House of Commons 26, 150, 195, 200, 204, 220-221
Hyndley, Lord 40, 44, 50, 53, 67, 84, 85, 90, 99, 100, 103, 106, 109, 112, 133

incomes policy 40, 59, 175, 179, 194-5, 231-3, 235-8
Industrial Charter (1947) 123

Jay, Douglas 45, 215
Jeger, George 129
Joint Consultative Committee 92-3
Joint National Negotiating Committee (JNNC) (*see* also conciliation machinery, wages) 36-7, 44, 84, 189, 191, 245, 247
Joint Production Committee 52, 63, 85
Joint Standing Consultative Committee 3, 17
Jones, Aubrey 130, 164-5, 166
Jones, Ernest 6, 7, 63, 70, 71, 85, 164, 189
Jones, W.E. 69, 71, 107, 134, 143, 152, 175, 178, 182, 187, 190, 199

Kane, Jock 228
Kane, Mick 178
Kelley, Dick 225

Kellher, Bill 143
Kent 43, 54, 82, 99, 175, 176, 181, 182, 216, 220, 231

Labour Department (NCB) 107-8, 152
Labour governments 11, 24,
  1945
  and coal industry 33-41
  and coal production 41-52
  and miners' wages 53-73
  1964
  1965 closure crisis 202-07, 214-9
  1967 closure crisis 219-29
  and coal policy 198-202, 248-50
  and nationalisation 21-6
  and nuclear power 239-4
  and wages 229-39
Lambton, Viscount 130
Lancashire (see also *North Western Area*) 5, 6, 42, 82, 85, 102, 142, 153, 167
Lancaster, Col. Claude 128, 130
Latham, T. 113
Lawther, Will 36, 38, 42, 43, 44, 54, 55, 57, 58, 61, 62, 63, 64, 65, 66, 67, 68, 69, 70, 71, 84, 90, 101, 107, 127, 132, 134, 137, 139, 140, 141, 152, 159, 197
Leathers, Lord 125, 128, 129, 131, 132, 134, 135, 141, 142, 144, 148, 153
Lee, Fred 201, 202, 203, 204, 206, 213, 214, 215, 216, 217, 240, 241, 242
Let Us Face the Future 28
Let's Go with Labour! 202
Llewellyn, D 91
Lloyd George, Gwilym 25, 26, 27
Lloyd, Geoffrey 110-112, 123, 124, 125, 126, 129, 130, 144, 146, 147, 148, 149, 150, 151, 153, 154, 160, 161, 164
Lloyd, Selwyn 192, 194

Machen, J.R.A. 110
MacLeod, Ian 166, 182
Macmillan, Harold 81, 125, 163, 175, 176, 182, 183, 185, 189, 191, 192, 194, 195, 196, 199, 201, 203
Main, Dick 178, 227

management 13-15, 18-20, 23-5, 27-8, 34, 37, 82, 84, 86-01, 104-5, 108, 115, 122-3, 128-9, 143, 145, 148-9, 151-9, 160, 162, 182, 187, 203, 206, 225, 243, 245-6,
Manchester Collieries 96
Marchwood power station 148-50
Marsh, Richard 216-23, 226, 226, 240-44
Mason, Roy 229, 242, 243-4
Mass Observation 92, 93, 94, 101, 103
Maud, Sir John 114
McKendrick, J. 159
Midlands, 14, 102, 106, 142, 146, 157, 159, 163, 177, 227
Mills, Sir Percy 184, 188, 189, 190
Miners and the Board 95
Miners Case, The 22
Miners' Parliamentary Group 36, 179, 203, 221-2, 225-6
Mineworkers Federation of Great Britain (MFGB) 1-2, 3-4, 6-7, 12, 17, 20-22, 24-5
Mining Association of Great Britain 11, 17, 20-21, 26
mining engineers 12, 13, 15
Ministerial Coal Committee 45
Ministry of Fuel and Power (*also* Ministry of Power) 11-2, 17, 19, 26-8, 34-5, 38, 42, 45, 87-9, 114, 124, 129, 131, 150-1, 179, 185, 194, 196-7, 215, 218-9, 223-4, 243-4
Ministry of Labour 35, 89-0, 138
Ministry of Power (*see* Ministry of Fuel and Power)
Moffat, Abe 21, 53, 55, 59, 61, 66, 70-71, 86, 100, 104-6, 160, 162, 205
Moffat, Alex 141, 181, 183, 186, 18, 191
Monckton, Sir Walter 124, 125, 127, 129, 135, 136, 138, 161, 162
Morrison, Herbert 23, 24, 25, 26, 28, 39, 47-8, 86, 91, 95, 155, 158,

Nabarro, Sir Gerald 128, 129, 131, 150
NACODS 178
National Board for Prices and Incomes (NBPI) 233, 234, 235
National Coal Board (NCB)
  1957-58 closures 184-97
  1965 closures 197-202, 213-9
  1967 closures 219-29

and coal policy 184-5, 188-9, 192-3, 198-06, 214-5, 219-29, 249-50
and industrial relations 34, 38, 40, 44, 60-61, 82-93, 102, 104, 108, 115, 132, 139, 145, 153, 158, 161, 177, 181-2, 189, 214, 245, 247
and production 39, 40, 41-52, 61-3, 67-8, 82-3, 85-6, 92-3, 101-3, 105-6, 145-7, 156-7, 159, 163, 179, 184, 186-9, 191-2, 199-205
and strikes 83-8, 101-3, 126-7, 131-5, 161-3, 226-8, 230-31
and wage bargaining 53-61, 88-9, 101-2, 108-9, 110-12, 131-7, 138-41, 145-7, 161-3, 176-9, 180-83, 193-5, 229-39
Conservative attitude to 15, 22-31, 150, 151-9
nature of 9, 22-7, 34-6, 38-9, 42-3, 81-2, 84, 86-7, 92-9, 105-7, 121-3, 129-1, 154-7, 159-0, 176-7, 179-3
resists nuclear power 218-9, 221, 223, 239-4
National Executive Committee (NUM), 6-9, 38, 42-4, 46-1, 53-8, 59-1, 63-8, 70-73, 82-4, 88, 91, 93, 96-7, 100, 103-7, 113-14, 129, 133, 137-43, 152, 158-3, 167, 175, 177-2, 185-2, 190, 192, 202-6, 215-6, 218-23, 225-7, 229-8, 246, 249-50
National Plan, The (1965) 201, 206, 213
National Power Loading Agreement (NPLA) 232, 233, 234, 236, 237, 238
National Reference Tribunal 44, 90, 108, 138-9, 160-61, 176-8, 181-2, 189
National Union of Mineworkers
  1944 constitution 4-6
  1957-58 closures 184-93
  1965 closure crisis 223-29
  1967 closure crisis 202-7, 214-9
  attitude towards coal prices 108-15
  attitude to Conservatives 121-2, 132-5, 137-44, 127-8
  Conservative coal policy 147-8, 151-4
  ethos 1-3, 93-108
  Fleck Report 152-9

formation of 2-3
internal politics 6-11, 61-73, 137-44
Labour's fuel policy 197-202
nationalisation 21-5, 26-9, 93-108
NCB 33-41, 82-93
nuclear power 239-44,
output 43-54,
political exclusion of 244-50
production 41-53
wages 53-61, 131-2, 175-84, 193-5, 229-39
National Wages Agreement (1944) 36
Nationalisation 11-12, 22-7, 34-6, 38-9, 42-3, 81-2, 86-7, 92-9, 105-7, 121-3, 127, 129-1, 154-7, 159-60, 176-7, 179-3, 234, 237, 245
NEDC 194, 195, 201
New Deal for Coal 206
Noel-Baker, Philip 51
North Derbyshire 90
North Sea gas 202, 217, 218, 219, 221, 223
North Western Area (*see* Lancashire)
Northumberland 4, 49, 59, 70, 72
Nottinghamshire 2-3, 4, 5, 11-12, 55, 59, 92, 133, 152, 157, 162, 163, 176, 187, 228, 238, 250
nuclear power 126, 128, 145, 150, 162, 165, 188, 202, 213, 218-9, 221, 223, 239-4

Oil 34, 113-4, 126, 128, 146-51, 162, 164-5, 179-80, 184-5, 187-8, 192, 199, 200, 202-4, 219-21, 228-9, 239, 242
Output per Manshift 12, 26

Panitch, Leo 248
Parkin, Herbert 190
pay (*see*, wages)
Pay Pause (1961) 194-5
Paynter, Bill 58, 61, 68, 69, 70, 71, 163, 176, 180, 182, 183, 189, 190, 192, 193, 194, 197, 205, 207, 215, 220, 222, 228, 229, 230, 237, 242, 249
Pearson, Willie 57, 86
Plan for Coal (CP) 67
Plan for Coal (NCB) 107, 166
piece rates (*see* National Power Loading Agreement) 59, 102, 140, 150, 163, 185
pit politics 7, 11, 37-8, 45, 49, 50, 53, 57, 82, 85-6, 91, 99, 102, 104, 106, 139, 149-50, 159-64, 177, 190, 205-6, 214, 219, 222, 230-231, 243, 250.
Pit Production Committee 83
Pollitt, Harry 67
Powell Duffryn 96
Pratt, Arthur 159, 163, 227-8
President (NUM) 10-11, 36, 67, 71, 84, 98, 152, 162, 164, 177, 187, 199, 204, 235
production 2-3, 7, 12-6, 25, 27, 35, 39, 40, 41-52, 53, 61-3, 65, 67-8, 70, 72, 82-3, 85-6, 92-3, 98, 101-3, 105-6, 108, 114, 129, 130, 145-7, 148, 150, 154, 156-7, 159, 163, 179, 184, 186-9, 191-2, 199-205, 218, 223
Productivity, Prices and Incomes (1965) 232-3
Prospects for 1961 192
Public Accounts Committee 35
PWR (*see* nuclear power) 240

Raikes, Sir Victor 129
Regional policy, 196-7
Reid Report 12-18, 19, 23, 27-8, 42, 82, 92
Reid, Sir Charles 12, 99-100
Reorganisation Sub-Committee 2
Revised Plan for Coal 198-9
Revisionism 137
Ridley Committee (1952) 184-5
Robens, Sir Alfred 190, 191, 192, 193, 194, 195, 196, 200, 205, 206, 207, 215, 216, 221, 222, 225, 226, 229, 239, 240, 241, 242, 243, 244
Robertson, A.J. 45
Robinson, Jack 230
Rossington 92
Rowe, Bill 216, 220
Rule 41 6, 72
Rule 43 6, 72-3
Rulebook (1944), 3-7, 11, 72-3, 90, 140, 163

Samuel Commission (1926) 27

Sankey Commission (1919) 27
Saturday Working Agreement 44, 47, 48, 49, 50, 133, 134, 141, 142, 143, 146, 147, 158, 161, 164, 176, 184
Savage, Bill 238
Scargill, Arthur 216, 221
Schofield, Sid 237-8
Schumacher, E.F. 199, 201, 205
Scotland 6, 21, 54, 55, 69, 71, 85, 108, 140, 141, 159, 162, 177, 181, 185, 196, 230, 251, 259,
Seaton Carew 221, 242-4
sectoral corporatism 34, 82, 167, 246
Select Committee on Nationalised Industries (SCNI) 113-15, 128
Selsdon Park Hotel 219, 224, 226
Shinwell, Emmanuel 27-8, 36, 37, 38, 39, 42, 43, 45, 46, 83, 89, 100, 121, 129, 219, 225, 240
Shore, Peter 222, 225
Snowden, Philip 23
Somerset 91
South Derbyshire 231
South Wales 42, 52, 58, 68, 69, 70, 71, 72, 85, 96, 99, 102, 127, 140, 141, 143, 160, 162, 163, 175, 176, 177, 180, 181, 192, 195, 196, 206, 231, 239, 250
Spencer, George 2, 3, 4, 5, 11
state capitalism 81
Street, Sir Arthur 103
strikes, 13-5, 53, 72, 83-8, 91, 101-3, 108, 121, 126-7, 131-5, 147, 149, 161-3, 177, 182, 189, 190, 226-8, 230-31
Suez Crisis 126, 177, 184, 239
Sunningdale Conference 229
surface polity 244, 245, 247, 248-50
surface workers 162-3, 180
Swain, Tom 214
Swingler, Stephen 124
Swinton, Lord 134

Taking Stock 39-40
Taylor, Sammy 183
Technical Advisory Committee (*see* Reid Report)
This Is the Road 123
Thorne 92

Thorneycroft, Peter 166
Tighe, Jack 187
Time for Decision 217
Trades Union Congress 21-2, 24-5, 27, 34, 37, 54-55, 57-9, 66, 68, 73, 83, 105, 126-7, 136, 138, 162, 176, 188, 192, 195, 197-8, 200, 207, 231-4, 235-6, 246
union politics 7, 11, 34-5, 47-9, 50-2, 53-4, 56-7, 82-92, 93-108, 131, 133-4, 137-44, 157-9, 161-2, 176-80, 189-91, 199, 202-7, 215-6, 218, 218, 222, 226-9, 230-31, 232-6, 243-4, 244-50
unofficial disputes (*see* strikes)

Vesting Day 38-9, 41, 44, 85, 86, 96, 101, 105

wages (*see* conciliation machinery, National Reference Tribunal, and JNNC) 9, 13-5, 18-9, 53-61, 67, 69, 84, 88-9, 101-2, 104, 106, 108-9, 110-2, 118, 122, 124, 126, 128, 131-7, 138-41, 145-7, 149, 156, 159, 161-3, 165, 167, 176-9, 180-3, 185, 189, 191, 193-5, 197, 229-39, 245, 250
wage structure (*see* National Power Loading Agreement)
Wainwright, Edwin 219-20, 222
Waleswood 90-91
Watkinson, Harold 141, 166
Watson, Jack 132
Watson, Sam 3, 55, 63, 66, 69, 70-71, 100, 125, 159, 160, 178, 188, 191, 231
Webber, Bill 221
West Midlands 102, 106
Williams, Glyn 70, 72, 177
Williams, Iestyn 85
Williams, Sir Evan 18
Wilson, Harold 46, 197, 198, 199, 200, 201, 202, 213, 217, 221, 222, 223, 225, 234, 235, 236, 239, 241, 242, 243, 244, 248, 249, 250
Wood, J. 140
Wood, Richard 185, 190, 192, 193, 195, 196
Woolton, Lord 136, 149, 162
Wynn, Bert 91, 133, 142, 160

Yorkshire 3, 5, 21, 59, 70, 83, 84, 87, 92, 94, 102, 104, 106, 110, 132, 137, 142, 143, 146, 162, 163, 177, 181, 182, 183, 189, 216, 218, 220, 227, 228, 230, 231, 237, 240, 250

Zweig, F 42, 87, 92, 102